Inquiry into College Mathematics

Inquiry into College Mathematics

MILTON D. EULENBERG
THEODORE S. SUNKO

Wilbur Wright College
Chicago, Illinois

John Wiley & Sons, Inc.
NEW YORK · LONDON · SYDNEY · TORONTO

Copyright © 1969, by John Wiley & Sons, Inc.

All Rights reserved. No part of this book may be reproduced by any means, nor transmitted, nor translated into a machine language without the written permission of the publisher.

Library of Congress Catalog Card Number: 79-77831
SBN 471 24725 1

Printed in the United States of America

Preface

The most perplexing problem encountered by the writer of a mathematics textbook for college freshmen is the selection of topics that will introduce the student to what is "modern" in mathematics but will not neglect those areas of classical mathematics which are still necessary for a *working* knowledge of mathematics. We are in agreement with the belief that, for college freshmen, the term "modern" should apply more to the spirit in which the book is written than to the choice of subject matter. The prevailing viewpoint of mathematics as the science of axiomatics and not limited to what is quantitative or measurable must not be overlooked. However, the tremendous development of mathematical science based on applications of the function concept, with its modern refinements, is still basic to an appreciation of the powerful interplay of mathematical ideas. Since the usefulness of mathematics has always been an important factor in its vitality as an area for thoughtful inquiry, several elementary functions are discussed in some detail.

We have tempered our axiomatic approach with the awareness that great ideas are frequently discovered and used long before they are proved by formal deduction. We have, therefore, given some recognition to the inductive approach. An inquiry into number systems is presented both for its intrinsic values and because it provides clear demonstrations of deductive processes in a nongeometric setting. Euclidean geometry has not been neglected; the concrete models of geometry serve to link the more difficult abstractions of axiomatic theory and formal logic with the reality (at least to the student) of point, line, and plane. Because many college freshmen have not mastered even the simple techniques of solving equations, we have included these topics in the chapters on linear relations and quadratic functions, though from a more mature point of view. Among the abuses of the "reforms" of mathematics is the excessive use of symbols which often obscure rather than clarify ideas, and we have used symbolism only when the purposes of clarity and exactness of expression are served.

Paramount in our minds is the belief that only through thought-provoking exercises, which must at times include a few problems that may be described as routine, can a student really *learn* what he is supposed to be studying. We have constructed the exercises with more than the usual amount of care, and they form an integral part of our book.

The experience of the authors has demonstrated that the contents of this book can be used to satisfy the needs of the following kinds of students.

1. Nonscience students who need a three or six semester hour course in mathematical concepts as an important component of a liberal education.

2. Prospective teachers and other specialized groups who need a three or six semester hour background course in mathematics.

It is suggested that a six semester hour course cover all the chapters in the order in which they are presented, although Chapters 1 to 4 and 8 and 9 are largely independent of one another and, therefore, may be varied in sequence. The special needs of the group under consideration will dictate the choice of topics for a three semester hour course. Such a course may be composed of the following combinations: Chapters 1 to 7; Chapters 1, 2, and 5 to 8; Chapters 1 and 3 to 8; or Chapters 2 to 7.

We wish to express our appreciation to the many people whose assistance and encouragement helped us in the development of this book. Particular acknowledgement is due to our wives for providing secretarial skills and exercising the virtues of patience, understanding, and encouragement, and to Professors Lawrence Balch, James Gray, Howard James, Thomas Jones, and Charlene Pappin of our department of Mathematics, for their critical assessment of the expository material and exercises.

We are hopeful that, with a dedicated teacher, the use of this book will encourage students to pursue further an inquiry into the ideas of mathematics.

Milton D. Eulenberg
Theodore S. Sunko

Contents

Chapter 1. WHAT IS MATHEMATICAL REASONING? ... 1

1. Inductive Inference ... 1
2. The Nature of Deductive Reasoning ... 5
3. Deductive Reasoning in a Logical System ... 14
4. Sentences, Symbols, and Truth Tables ... 22
5. The Conditional Sentence ... 27
6. Valid Arguments and Tautologies ... 34
7. The Concept of Set ... 39
8. Related Sets ... 46

Chapter 2. WHAT IS A NUMBER SYSTEM? ... 53

9. The Development of a System of Numeration ... 53
10. Numeration for Computers ... 61
11. Mathematical Systems ... 66
12. Peano Axioms and the Natural Numbers ... 74
13. The System of Integers ... 81
14. The Rational Numbers ... 89
15. The Real Number System ... 94

Chapter 3. WHAT IS THE AXIOMATIC BASIS OF ALGEBRA? ... 103

16. The Equality Relation ... 103
17. Field Axioms ... 107

Contents

18. Algebraic Expressions and Operations	110
19. Equations and Inequalities in One Variable	119
20. Applications of Simple Linear Relations	123

Chapter 4. WHAT ARE SOME OF THE FUNDAMENTAL PROPERTIES OF GEOMETRIC FORMS? 128

21. The Beginnings of Geometry	128
22. The Elements of Geometry	132
23. Polygons and the Circle	137
24. Congruence and Similarity	144
25. Geometric Forms in Space	153
26. Symmetry	160

Chapter 5. WHAT IS A FUNCTION? 167

27. The Function Concept	167
28. Tabular Representation	174
29. Rectangular Coordinate System	178
30. Graphic Representation	182
31. Functions and Change	188

Chapter 6. WHAT IS A LINEAR FUNCTION? 195

32. The Linear Function	195
33. Rate of Change of Linear Functions	200
34. Systems of Linear Functions	204
35. Applications of Linear Systems	211

Chapter 7. WHAT IS A QUADRATIC FUNCTION? 215

36. The Quadratic Function	215
37. The Rate of Change of the Quadratic Function	221
38. Quadratic Equations and Their Solutions	228
39. Extending the Use of the Quadratic Formula	233
40. Applications of the Quadratic Function	240

Chapter 8. HOW ARE STATISTICS USED TO PRESENT AND MEASURE DATA?   243

41. Uses of Statistics   243
42. Organization of Data   246
43. Measures of Central Tendency   252
44. Measures of Dispersion   261
45. Frequency Distributions and the Normal Curve   268

Chapter 9. WHAT IS MODERN MATHEMATICS?   275

46. Characteristics of Modern Mathematics   275
47. Boolean Algebra   276
48. Linear Programming   283
49. Finite Geometry   288
50. Transfinite Numbers   294

APPENDIX   303

ANSWERS TO ODD-NUMBERED PROBLEMS   305–325

INDEX   327–331

Inquiry into College Mathematics

1

What is Mathematical Reasoning?

1. INDUCTIVE INFERENCE

Some students may find it surprising that a textbook of mathematics should begin with a chapter on ideas which seem to be nonmathematical. Indeed, the first question a student frequently asks when he registers for his first college course in mathematics is, "Is it basically algebra or geometry?" To him mathematics must be related to numbers or geometric figures. The modern viewpoint does not neglect these aspects of mathematics but emphasizes mathematics as a science which draws conclusions from certain basic assumptions, that is, a science of reasoning. While much of mathematics does concern itself with numbers and their properties, we shall use nonnumerical examples whenever they serve to illustrate the reasoning process.

Inductive Reasoning

A very common type of advertisement states, "This book has helped thousands to increase their earning power." The reader, it is hoped, will infer that since the book has helped thousands, it will also help him. What are the elements of this thought process? On the basis of thousands of observations or experiences, a conclusion has been reached. What is true of thousands of people must be true of all people. The old saying, "A burned child dreads the fire," is another example of a conclusion reached, in this case, from a single experience. Since the child has been hurt by one fire, *all* fires are harmful. We call this process of drawing conclusions from experience or observation, *inductive reasoning*.

The Nature of Inductive Reasoning

We hasten to state that inductive reasoning does not always result in a true conclusion. A young man, finding that the first ten girls he asks to the school dance have already been dated for that occasion, may conclude that it is useless to ask any other girls; they will all refuse. If he asks one more girl and she accepts his invitation, his conclusion is incorrect. Similarly, if a child lost in the woods in subzero weather is able to kindle a fire, he may well be saved from freezing to death. All fires are not harmful, if properly used. We see that a single result contrary to the conclusion reached by previous experience is enough to render the conclusion untrue. It follows that certain conclusions reached by inductive reasoning are tenuous—for example, even though in our total experience no one has ever counted to one million, we cannot say that no one ever will.

On the other hand, the process of inductive reasoning is indispensable to scientific research. The report of the Surgeon General that overwhelming evidence indicates that smoking is a major cause of lung cancer is taken seriously by many people, despite the fact that there is no actual proof that this is true. The discovery of penicillin was the result of an observation that a certain mold destroyed bacteria, and thousands of lives have been saved by this drug. Inductive reasoning is a powerful means of *discovery*, and the fact that observations sometimes lead to erroneous conclusions simply means that care must be exercised in arriving at conclusions in this manner.

Two further examples will serve to caution the reader that conclusions reached through inductive reasoning may or may not be true.

EXAMPLE 1. By trial, it is found that the numbers 4, 24, 44, and 64 are all divisible by 2. Therefore we conclude that all numbers ending in 4 are divisible by 2. This conclusion happens to be true for our system of whole numbers.

EXAMPLE 2. By trial, we also find that the numbers 4, 24, 44, and 64 are all divisible by 4. Therefore we conclude that all numbers ending in 4 are divisible by 4. A single contrary example, the number 34, which is not divisible by 4, disproves our conclusion.

In Example 1 we stated that the conclusion *happens to be true*. Thus we emphasize the fact that in inductive reasoning no amount of observation proves a conclusion true, but a single contrary result is all that is needed to prove the conclusion false.

EXERCISE 1

INDUCTIVE REASONING

In Problems 1 to 9, indicate whether you agree or do not agree with the stated conclusions.

1. A study of the case history of diabetics reveals that most of them complained of excessive thirst and, therefore, drank large amounts of water. The conclusion is that drinking large amounts of water is a major factor in causing diabetes.
2. A student observes that whenever he measures the five angles of a pentagon, the sum of the five angles is very close to 540 degrees. He concludes that the sum of the angles of *every* pentagon is 540 degrees.
3. The ancient Greeks noticed that the ratio of the circumference of any circle to the diameter of the circle was always a little more than 3 to 1. Therefore, although they did not know the exact value of this ratio, they concluded that it was the same for all circles and they called it π.
4. A bettor at a race track noticed that he won all of six bets he placed on horses whose names began with B. Therefore he concluded that he would always win when he bet on a horse whose name began with B.
5. Every time John added two odd numbers, the sum was an even number. He concluded that the sum of two odd numbers is always an even number.
6. John (Problem 5) also concluded that since the sum of two odd numbers is always an even number, the sum of three or more odd numbers would also be an even number.
7. It is observed that the product of two, three, or four odd numbers is always an odd number. Therefore the product of any number of odd numbers is always an odd number.
8. A tile setter found that all of the tiles he worked with were either squares or hexagons. He concluded that only square or hexagonal tiles can be used as floor tiles.
9. A student in mechanical drawing found that when he drew six consecutive chords in a circle, each equal in length to the radius of the circle, a regular hexagon was formed. He concluded that this would be true for every circle he drew.
10. By drawing all possible diagonals of polygons of 4, 5, 6, 7, and 8 sides, find, by inductive reasoning, a formula relating the number of diagonals of a polygon to the number of sides of the polygon.
11. The squares of the numbers 2, 3, 4, 5, 6, 7, 8, 9, 10, namely, 4, 9, 16, etc., are all greater than the numbers. Thus $4 > 2$, $9 > 3$, $16 > 4$, and so on. Is the conclusion that the square of any number is greater than the number itself true for all numbers? Remember that a single contrary example will serve to disprove the conclusion.

12. Observe the triads of consecutive odd numbers: 3, 5, 7; 5, 7, 9; 7, 9, 11; 9, 11, 13; and so on. Can you find a triad where none of the numbers is divisible by 3? Express your result in the form of a statement and a conclusion.

13. Note that $2 \times 2 = 4$ and $2 + 2 = 4$. Note also that $3 + 2 = 5$, but $3 \times 2 = 6$; $4 + 3 = 7$ but $4 \times 3 = 12$. May we conclude that the only case where the sum of two numbers equals their product is when both numbers are 2? Do not restrict yourself to whole numbers.

14. A child notices that in every long-division problem he does, the remainder is less than the divisor. He concludes that this will always be true when the division is properly done. Is he correct?

15. A student knows that when the sum of the digits of a number is divisible by 3, the number itself is divisible by 3. For example, since the sum of the digits of 3471 is $3 + 4 + 7 + 1$, or 15, which is divisible by 3, it is certain that the number itself, 3471, is also divisible by 3. He concludes that when the sum of the digits of a number is divisible by 4, the number itself is divisible by 4, and so on. Is his conclusion true?

16. A student substitutes 1, 2, and 3, for x, in the expression $x^3 - 6x^2 + 11x - 6$, and finds that in each case the result equals zero. He concludes that this expression equals zero when any number is substituted for x. Is he correct?

17. The student of Problem 16 also observes that when he substitutes 1, 2, and 3 in the expression $x^2 + 2x + 1$, the result in each case is a perfect square, that is, 4, 9, and 16. Is he justified in concluding that the expression will be a perfect square for every number he chooses for x?

18. The numbers 1, 3, 6, 10, 15, 21, and so on, are called triangular numbers. They are formed by the sums:
$$1 + 2 = 3$$
$$1 + 2 + 3 = 6$$
$$1 + 2 + 3 + 4 = 10$$
$$1 + 2 + 3 + 4 + 5 = 15$$
$$1 + 2 + 3 + 4 + 5 + 6 = 21$$

By inductive reasoning, what can be concluded about the sum of certain pairs of these triangular numbers?

19. Select any two unequal positive numbers a and b. Form the sum $S = a^2 + b^2$ and the product $P = 2ab$. Thus if $a = 2$ and $b = 3$, $S = a^2 + b^2 = 13$, and $P = 2ab = 12$. Can you reach any conclusions concerning $S + P$ and $S - P$?

20. How is the conclusion of Problem 19 changed if two equal numbers are substituted for a and b?

21. Study the equalities:
$$3^2 - 1 = 8$$
$$5^2 - 1 = 24$$
$$7^2 - 1 = 48$$
What conclusion can you draw?

22. Find the product of any four consecutive numbers and add 1 to this product. What do you conclude?

23. Observe the following number pattern:

$$1^3 = 1$$
$$2^3 = 3 + 5$$
$$3^3 = 7 + 9 + 11$$
$$4^3 = 13 + 15 + 17 + 19$$
$$5^3 = 21 + 23 + 25 + 27 + 29$$

(a) Express 6^3, 7^3, 8^3, and 9^3 in a similar way.
(b) By inductive reasoning what can be concluded about the cube of any number?

24. Select any whole number n greater than 1, and then form the numbers a, b, and c, using the relations:

$$a = n^2 - 1$$
$$b = 2n$$
$$c = n^2 + 1$$

Substitute several other numbers for n where, again, n is greater than 1. Do the numbers a, b, and c, obtained for each substitution for n, have a familiar relationship?

2. THE NATURE OF DEDUCTIVE REASONING

As we have seen, inductive reasoning is a thought process whereby repeated observations lead to a conclusion but do not prove the conclusion. We may observe 10,000 triangles and find that the sum of the angles of each triangle is 180 degrees, but we have no proof that this is true of all triangles. The reader may recall that this fact was proved in his high school geometry course by using certain basic assumptions and previously proved theorems. When such assumptions lead to an inevitable conclusion, we say that the conclusion has been reached by *deductive reasoning*. Deductive reasoning needs no specific observations to support a conclusion, but observation certainly plays an important part in discovering, or suggesting, a conclusion. It is unfortunate that the typical deductive proof in a textbook of mathematics gives no evidence of the inductive reasoning that may have suggested the conclusion. For example, it is quite likely that observation played a great part in suggesting that the sum of the angles of any triangle equals 180 degrees, but the deductive proof shows no evidence of this. The fact that the square on the hypotenuse of a right triangle is equal to the sum of the squares on the two legs was known to the Babylonians more than a thousand years before Pythagoras, but to Pythagoras goes the credit for being the first to *prove* it.

What is Mathematical Reasoning?

The Form of a Deductive Proof

Let us give a simple example of deductive reasoning.

> All artists are starving.
> John is an artist.
> Therefore, John is starving.

What are the elements of this thought process? Two assumptions are made. The first statement, that all artists are starving, may or may not be true, but we *assume* that it is true. Similarly we accept as true the second statement, that John is an artist, even though we may not know who John is. Then, on the basis of these two *assumptions*, we are *forced* to conclude that John is starving. There is no alternative and, therefore, we say that we have *proved* the conclusion.

The Syllogism

The particular form of deductive reasoning we have used here is known as *a syllogism*. A classical example of deductive reasoning is the famous syllogism of Aristotle.

> All men are mortal.
> Socrates is a man.
> Therefore Socrates is mortal.

We may generalize the form of a syllogism as:

> All A is B.
> All C is A.
> Therefore all C is B.

A common error made by students is illustrated in the following example.

> All men are mortal.
> All women are mortal.
> Therefore all women are men.

The error lies in the use of an incorrect form of reasoning.

> All A is B.
> All C is B.
> Therefore all C is A.

While the conclusion, "Therefore all women are men," is obviously ridiculous, a similar error may appear in more subtle form, as illustrated in the following example.

> 27 is an odd number.
> The third power of an odd number is an odd number.
> Therefore 27 is the third power of an odd number.

Although it is true that 27 is the third power of an odd number ($27 = 3^3$), this fact does not follow from the first two statements. The error is easily seen if we substitute 25 for 27, since 25 is not the third power of an odd number. Therefore there is clearly an error in the reasoning process.

Validity and Truth of a Deductive Argument[1]

The preceding examples illustrate two properties of a deductive argument, *validity* and *truth*. A deductive argument is valid if the conclusion follows from the assumptions made. The syllogism:

> Green things are twenty feet high.
> Horses are green.
> Therefore horses are twenty feet high.

is quite valid, although the assumptions as well as the conclusion are nonsense.

Thus we see that the validity of an argument refers to the form of the argument and has nothing to do with the truth of either the assumptions or the conclusion. The conclusion that 27 is the third power of an odd number was true but not valid, since it did not follow from the assumptions made.

From the point of view of producing a logical argument, we are interested only in its validity; have we reasoned correctly from the given assumptions? From a practical viewpoint, however, we are also concerned with the truth of the conclusion. Consider the following argument.

> Fair teachers do not fail students.
> My teacher is fair.
> Therefore my teacher will not fail me.

Aside from the various meanings of the word "fair," there is grave doubt that the first assumption is reasonable or true. A student who selected this teacher with the idea that he could loaf through the course would be due for a surprise when he received his grade at the end of the semester. The argument is valid, since the reasoning process is correct, but the conclusion is certainly unlikely to be true.

Note that the general form of the syllogism previously stated:

> All A is B.
> All C is A.
> Therefore all C is B.

serves to further illustrate the nature of validity. Although we have no knowledge of what A, B, and C may be, the argument is valid for whatever they are. Whether or not the conclusion is *true* is another matter. If both assumptions are true, the conclusion is true. If either of these statements is false, the conclusion may or may not be true, but the conclusion is valid.

---

[1] The term "argument" refers to the process whereby the conclusion is reached.

What is Mathematical Reasoning?

The Use of Diagrams in Deductive Reasoning

The use of diagrams, as illustrated in the following examples, is quite helpful in arriving at a conclusion or in showing that there is no conclusion.

EXAMPLE 1
All French women are beautiful.
Marie is a French woman.
Therefore Marie is beautiful.

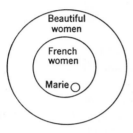

Figure 1.1

Solution

In Figure 1.1, we first draw the outer circle to include the set[2] of all beautiful women.

The second circle is drawn entirely within the first to show that the set of all French women is a part of the set of beautiful women.

Since Marie is French, the circle representing her is drawn inside the circle of French women.

Since the circle representing Marie is also inside the circle of beautiful women, it follows that Marie is beautiful.

EXAMPLE 2
All French women are beautiful.
Marie is beautiful.
No conclusion.

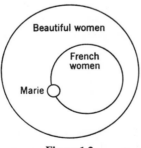

Figure 1.2

Solution

In Figure 1.2 the outer circle is drawn to include all beautiful women.

The inner circle is drawn entirely within the circle of beautiful women, since all French women are beautiful.

Since Marie is beautiful, we are *compelled* to put the circle representing Marie inside the circle of beautiful women. Although the assumptions do not compel us to put Marie in the circle of French women, we draw the circle representing Marie so as to indicate that she might be French.

There is no conclusion. Although Marie may be French, she is not necessarily French; only necessary conclusions are considered. *Unless a conclusion is inescapable, the argument is not valid.*

---

[2] The term *set* is used here in the ordinary sense, as in a set of dishes or a set of books. In Section 7 the particular mathematical meaning of *set* will be discussed.

2. The Nature of Deductive Reasoning

EXAMPLE 3
All beautiful women are French
Marie is beautiful.
Therefore Marie is French.

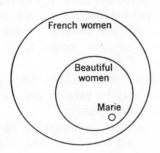

Figure 1.3

Solution

In Figure 1.3 the outer circle includes all French women.

The circle of beautiful women is drawn entirely within the outer circle, since all beautiful women are French.

The circle representing Marie is drawn inside the circle of beautiful women because of the second assumption.

Since Marie is also in the circle of French women, we must conclude that Marie is French.

The reader should carefully note the differences in the diagram for *all French women are beautiful* (Figure 1.2) and *all beautiful women are French* (Figure 1.3).

EXAMPLE 4
All beautiful women are French.
Marie is French.
No conclusion.

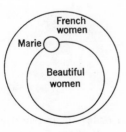

Figure 1.4

Solution

In Figure 1.4 the circles are drawn as in Figure 1.3 to show that beautiful women must be French.

The second assumption *compels* us to put the dot representing Marie inside the circle of French women. Although the assumptions do not compel the inclusion of Marie in the circle of beautiful women, the circle representing Marie is drawn to indicate that she might be beautiful.

There is no conclusion. Although Marie may be beautiful, she is not *necessarily* beautiful; as in Example 2, only necessary conclusions are considered.

10   What is Mathematical Reasoning?

EXAMPLE 5
Some beautiful women are French.
Marie is beautiful.
No conclusion.

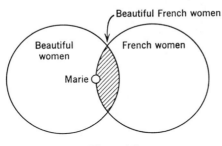

Figure 1.5

Solution

In Figure 1.5 the two circles overlap to show that only some beautiful women are French.

The second assumption compels us only to place the circle representing Marie in the circle of beautiful women, but her circle is drawn to indicate that she could also be French.

There is no conclusion. As in Example 4, Marie may be French, but she is not necessarily French.

If the second assumption were "Marie is French," a similar situation would prevail and no conclusion could be reached.

EXAMPLE 6
No beautiful women are French.
Marie is beautiful.
Therefore Marie is not French.

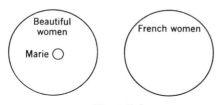

Figure 1.6

Solution

In Figure 1.6 the circle of beautiful women and the circle of French women are drawn entirely apart.

Therefore we must conclude that Marie is not French.

Compound Syllogisms

When more than two assumptions are made in order to arrive at a conclusion, we shall call the statements a compound syllogism. The following example was created by Lewis Carrol (C. L. Dodgson),[3] the author of *Alice in Wonderland*. The three assumptions are:

       1. Babies are illogical.
       2. Nobody is despised who can manage a crocodile.
       3. Illogical persons are despised.

---

[3] *Symbolic Logic and the Game of Logic*, Lewis Carroll, Dover, 1958, p. 112.

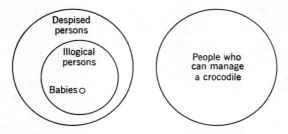

Figure 1.7

From statements 3 and 1, we may conclude:

4. Therefore babies are despised.

Statements 2 and 4 lead to the final conclusion:

Therefore babies cannot manage crocodiles.

The diagrams, as shown in Figure 1.7, clearly point the way to the conclusion, since babies are outside the circle of people who can manage crocodiles.

It is instructive to note that we may combine the three statements in another order and arrive at the same conclusion. Statements 2 and 3 can be combined to reach the conclusion, "Illogical people cannot manage crocodiles." Combining this conclusion with Statement 1, we reach the conclusion, as before, "Babies cannot manage crocodiles."

EXERCISE 2

Discuss the following syllogisms with respect to validity and (within your experience) truth. Use diagrams to illustrate your answers.

1. The sum of two even numbers is an even number.
   The numbers a and b are even numbers.
   Therefore the sum $a + b$ is an even number.
2. The sum of two even numbers is an even number.
   The sum $a + b$ is an even number.
   Therefore a and b are even numbers.
3. The product of two negative numbers is a positive number.
   The product $a \times b$ is a positive number.
   Therefore a and b are negative numbers.
4. The product of two negative numbers is a positive number.
   The numbers a and b are negative numbers.
   Therefore the product $a \times b$ is a positive number.

5. The product of two even numbers is an even number.
   The product $a \times b$ is an even number.
   Therefore a and b are even numbers.
6. The product of two even numbers is an even number.
   The numbers a and b are even numbers.
   Therefore the product $a \times b$ is an even number.
7. All numbers ending in 2 are even numbers.
   17 does not end in 2.
   Therefore 17 is not an even number.
8. All numbers ending in 2 are even numbers.
   The number 3412 ends in 2.
   Therefore 3412 is an even number.
9. All numbers ending in 2 are even numbers.
   24 does not end in 2.
   Therefore 24 is not an even number.
10. The fourth power of an even number is an even number.
    16 is an even number.
    Therefore 16 is the fourth power of an even number.
11. The fourth power of an even number is an even number.
    x is the fourth power of an even number.
    Therefore x is an even number.
12. The diagonal of a parallelogram divides it into two congruent triangles.
    This four-sided figure is divided by a diagonal into two congruent triangles.
    Therefore this figure is a parallelogram.
13. Many composers died penniless.
    Stephen Foster died penniless.
    Therefore Stephen Foster was a composer.
14. A barking dog never bites.
    My dog barks.
    Therefore my dog never bites.
15. A barking dog always bites.
    My dog always barks.
    Therefore my dog always bites.
16. Barking dogs never bite.
    My dog does not bark.
    Therefore my dog never bites.
17. A barking dog never bites.
    My dog never bites.
    Therefore my dog is always barking.

In the following problems try to draw a conclusion from the given statements. Use a diagram to illustrate your conclusion or lack of a conclusion. Remember that information other than that given may not be used.

2. *The Nature of Deductive Reasoning* 13

18. Writers are often brilliant people.
    Edgar Allen Poe was a brilliant person.
    Conclusion:
19. Writers of mystery stories have a great deal of imagination.
    Smith has written many mystery stories.
    Conclusion:
20. All bus drivers are skillful drivers.
    Jones is a bus driver.
    Conclusion:
21. People with colds use Kleenex.
    Mrs. Smith is using Kleenex.
    Conclusion:
22. Railroad engineers carry very accurate watches.
    Mr. Brown carries a very accurate watch.
    Conclusion:
23. When it rains the streets are very wet.
    The streets are very wet.
    Conclusion:
24. People who are over 21 years are adults.
    Mary is over 21 years of age.
    Conclusion:
25. Several presidents of the United States had beards.
    Abraham Lincoln had a beard.
    Conclusion:
26. Generals of the army have sometimes become presidents.
    Dwight Eisenhower was a general of the army.
    Conclusion:
27. People over 65 years of age are eligible for pensions.
    John is eligible for a pension.
    Conclusion:
28. Mine owners are wealthy people.
    Mr. Smith is wealthy.
    Conclusion:
29. To be called an adult you must be over 21 years of age.
    Mary is an adult.
    Conclusion:

In Problems 30 to 34, use all three statements to draw a conclusion.

30. Some laws are complicated.
    Complicated laws are confusing.
    Confusing laws are unsatisfactory.
    Conclusion:

14 *What is Mathematical Reasoning?*

31. Some problems are difficult.
    Difficult problems are interesting.
    Interesting problems are worthwhile.
    Conclusion:
32. All squares are quadrilaterals.
    All quadrilaterals are polygons.
    All polygons are bounded by straight lines.
    Conclusions:
33. Some students do their homework daily.
    Students who do their homework daily pass their examinations.
    Students who pass their examinations eventually graduate.
    Conclusion:
34. Most people have troubles in their lives.
    Troubles tend to develop strength of character.
    Strength of character brings happiness.
    Conclusion:

3. DEDUCTIVE REASONING IN A LOGICAL SYSTEM

In Section 2 the examples and excercises in deductive reasoning were all unrelated to one another. The conclusion, for example, "Therefore Socrates is mortal," was not used to arrive at any other conclusion. The reader will no doubt recall that in the study of geometry almost every conclusion (or theorem, as it was called) was used to prove another theorem. In this manner a whole system of theorems was developed, each theorem depending on one or more that were previously proved. The set of theorems constituted the science of geometry. This sequence of theorems is not peculiar to geometry alone but is present in every branch of mathematics and every other science. The succession of theorems is more easily seen, however, in the typical geometry textbook. It is with the development of a logical system that we now concern ourselves, taking the expression "logical system" to mean simply an orderly process of reasoning.

The Elements of a Deductive System

Looking again at the syllogism of Aristotle, it is clear that there are certain elements present in every deductive argument. These are defined terms, such as "mortal," assumptions (or axioms), such as "All men are mortal," and a conclusion, such as "Socrates is mortal." Not quite so evident but basic to every science is a selected group of undefined terms whose existence becomes evident whenever a definition is traced sufficiently far, as we shall see in the following paragraph. Thus the elements of a deductive science are the undefined terms, the defined terms, the axioms, and the conclusions.

Undefined Terms

Since all words must be defined in terms of other words, it is clear that we can never define all words. Perhaps the best illustration of the futility of trying to define all words is to consider a person, who, without any knowledge of French, attempts to find the meaning of a French word using only a French dictionary. Suppose the word is *herboriser*. In his French dictionary he finds:

herboriser = cueillir les plantes pour les étudier

Since the definition is entirely in French, it has no more meaning to the person than the word itself. If he continues to use the French dictionary to look up the words in the definition, he will be in the same predicament; with the same dictionary his search for the meaning of *herboriser* will be endless. To a lesser extent, we are faced with a similar situation in our own language. We find in a dictionary for college students the word "frantic" defined as "excited," "excited" as "agitated," "agitated" as "perturbed," "perturbed" as "disturbed"; eventually we may be led back to the word with which we started. Unless somewhere in this chain of definitions we find a word whose meaning we understand, we are bound to fail.

Since we cannot define all words, we choose some words or terms to remain undefined, hoping that their meaning will be the same to all who use them in a given science.[4] The words "point" and "set" as used in mathematics are undefined; their meaning is illustrated rather than defined. While the word "set" means one thing to a movie star, another to a beauty operator, and still another to a woodworker using glue, it has an entirely different meaning to a mathematician, and all mathematicians agree on its meaning in their science.

Defined Terms

On the basis of those words selected to remain undefined, we may define other words. The importance of definitions is strikingly illustrated in the conclusion reached (erroneously) in the last section: "All women are men." Although we referred to this conclusion as "obviously ridiculous," need it be? What do we mean by the word "men"? No one seriously doubts that the writers of the Declaration of Independence, in stating that "All men are

---

[4] Undefined words are *not* words without meaning; in fact most undefined words have meanings that are quite clear.

created equal," did not mean to exclude women from their right of equality. In this particular context the word "men" is intended to include women; as someone has said facetiously, "Man embraces woman." On the other hand, a "men wanted" sign posted on the door of a foundry or steel mill refers specifically to the male sex. Unless we know exactly what meaning is intended for the terms in a deductive argument, the validity as well as the conclusion of our argument remains in doubt.

The Characteristics of a Good Definition

Since definitions play an important part in deductive reasoning, we give some of the characteristics of a good definition.

1. The term to be defined should be placed in a particular class, and the properties that distinguish it from other members of the same class should be stated. Thus, to define the word "parallelogram" we say: *a parallelogram is a quadrilateral with opposite sides parallel*. The parallelogram then belongs to the class of four-sided figures, with the special property that its opposite sides are parallel. Note that the word "opposite" is not defined; in geometry it may remain an undefined term.

2. A definition should give no more information than is necessary to identify the term being defined, that is, it should not be redundant. The definition of a parallelogram as a quadrilateral with opposite sides parallel and equal is redundant. While it is true that the opposite sides of a parallelogram are equal, the equality of the opposite sides is a *consequence* of their being parallel and, therefore, is not an essential part of the definition.

3. A definition should not be circular. If a dollar is defined as being the equivalent of one hundred cents, and then a cent is defined as the equivalent of one-hundredth of a dollar, the value of both the dollar and the cent are still unknown. Similarly, if the word "indolent" is defined as "lazy" and the word "lazy" is defined as "indolent," there is still no clue to the meaning of either word.

4. A definition should be unambiguous, that is, it should have only one meaning. We see the fallacy of an ambiguous definition in the following syllogism.

>Calculating persons have few friends.
>Mathematicians make many calculations.
>Therefore mathematicians have few friends.

5. A definition is reversible. If the statement, "People over 21 years of age are adults," is the definition of an adult, then a person over 21 years of age is an adult, and an adult is a person over 21 years of age. On the other hand, the statement, "People over 21 years of age are sensible," is not a definition of

the word "sensible," and we cannot state that a sensible person is over 21 years of age.

Special Definitions of Mathematics

In mathematics, not only are terms defined but certain expressions and operations are also stated as definitions. Consider the familiar expression:

$$b^0 = 1$$

It is sometimes difficult to convince a student that this is a definition and is, therefore, *not proved*. Since $b^1 = b, b^2 = b \times b, b^3 = b \times b \times b$, and so on, it would not be unreasonable to decide that $b^0 = 0$. But a strong consideration in deciding what meaning to give to b^0 is the preservation of the basic laws of exponents. We shall see later that this requires us to *assign* the value of 1 to b^0, which is simply saying that we *define* b^0 as to equal 1.[5] We see, therefore, that definitions are not entirely arbitrary, for while we have the right to assign any value to b^0, the original definition of exponents for positive integers prescribes our choice if our definition of b^0 is to be useful.

Axioms, or Basic Assumptions

Just as we cannot define all terms in a logical system, we cannot prove all assertions we make. We see that if Theorem 5 is proved by means of Theorem 4, Theorem 4 by means of Theorem 3, and so on, we cannot go back indefinitely. What theorem shall we use to prove Theorem 1? The statements leading to the conclusion of Theorem 1 cannot be other theorems and, therefore, they must be the basic assumptions, or axioms, of our logical system. The axioms of a deductive system are perhaps more troublesome than the other elements. Being basic assumptions, their truth is immaterial as far as the validity of the conclusions is concerned. On the other hand, while mathematics is no longer regarded as a "tool subject" useful only for its applications to other sciences or fields of human endeavor, we cannot ignore the fact that applications will continue to be a major purpose and inspiration for study and research. We may pursue at great length, for example, the practically useless conjecture of Goldbach that every even number greater than 4 can be expressed as the sum of two prime numbers,[6] but no one can say that this will always remain without any practical application. What we are saying here is that our assumptions cannot ignore the basic facts of nature as we see them. For purely logical purposes we may

---

[5] More specifically, $b^0 = 1, b \neq 0$.
[6] Examples are: $28 = 11 + 17; 36 = 17 + 19; 14 = 3 + 11$.

assume (falsely) that the sum of two odd numbers is an odd number and arrive at many valid conclusions, but such an assumption would be foolish in the light of our daily experience, and the conclusions would be pointless. On the other hand, we no longer talk about self-evident truths, as axioms were once defined, because what is evident to one person may not be evident to another, and because we have no way of making any infallible decision as to the truth or falsity of an assumption.

The Desirable Characteristics of a Set of Axioms

We are concerned not only with individual axioms but with their relationship to one another and to the logical system that they build. We list here a number of properties that are desirable in a set of axioms.

1. A set of axioms should be consistent, that is, they should not lead to contradictory conclusions. This is perhaps the most important property that a set of axioms must possess.

2. A set of axioms should be independent, that is, it should not be possible to derive one axiom from another.

3. Axioms should be as few in number as possible. While this property is desirable, it is not essential.

4. A set of axioms should be fertile, that is, at least one theorem should be produced from them.

To illustrate these principles, let us consider the following problem. On the eve of graduation the principal of a high school discovers that he has lost the notes concerning the four top students and their respective ranks. He hastily summons a group of teachers who volunteer the following information.

1. I know that Allen was either second or third.
2. I remember that Charles was first or second.
3. I think that Bob and Dave were both ahead of Allen.

On the basis of statements 1 and 3, Allen would have third place and Bob and Dave would occupy first and second place. This obviously conflicts with statement 2 and, if we consider these statements as axioms, we have failed to satisfy the condition that they be consistent.

Since the third teacher is now not sure of his statement, he is replaced by two other teachers. The new information is then:

1. I know that Allen was either second or third.
2. I remember that Charles was either first or second.
3. I am sure that Bob ranked just below Charles.
4. I know that Dave was the lowest of the four.

The first three statements show that Allen was third, Charles was first, and Bob was second. Statement 4 is then unnecessary, since it follows from the first three. Clearly the first three statements, if they are accepted as true, are sufficient to solve the problem. They may be considered as a set of axioms which illustrate some properties of axioms we have discussed; they are minimal in number, consistent, and independent.

Informal Deductive Proof

When the elements of a deductive system are explicitly stated in order to arrive at a conclusion, the procedure is called a formal proof. Examples of formal proofs will be developed as an intrinsic part of the subsequent discussions in this book. In mathematical proofs, as well as in ordinary argumentation, however, it is often convenient to use what we shall call *informal deduction*, where the assumptions and other elements of a deductive proof are not explicitly stated. The following examples of deductive proofs will illustrate the nature of informal deduction and will further illustrate how deductive conclusions are frequently suggested by observation, that is, by inductive reasoning.

Consider an arbitrary array of sequences of three consecutive numbers, such as:

$$5, 6, 7$$
$$13, 14, 15$$
$$21, 22, 23$$
$$79, 80, 81$$

Is there any property common to all of these sets? With a little reflection, we might discover that in each, one of the numbers is divisible by 3. Further observation of other sets of three consecutive numbers will lead to the conclusion, reached inductively, that this is true for all such sets. We now attempt to *prove* that this is true by informal deduction, establishing the result as a theorem for further application. The reader may supply the reason for each of the statements.

Theorem 1. Given any three consecutive numbers, one of them is divisible by 3.

1. Let any three consecutive numbers be represented by k, $k + 1$, and $k + 2$.
2. If k is not divisible by 3, then the remainder is either 1 or 2.
3. If the remainder is 1, then by adding 2 to k, the number will be divisible by 3. If the remainder is 2, then by adding 1 to k, the number will be divisible by 3.
4. Therefore, one of the numbers, k, $k + 1$, or $k + 2$, is divisible by 3.

Thus we established, by informal deduction, a conclusion which was suggested by inductive reasoning.

We now consider several sequences of three consecutive odd numbers, such as:

$$7, 9, 11$$
$$27, 29, 31$$
$$45, 47, 49$$
$$79, 81, 83$$

We find, as before, that in each of these sequences of odd numbers, one number is divisible by 3. By informal deduction we shall try to prove our conjecture that every sequence of three consecutive odd numbers contains one number that is divisible by 3. In particular, we shall indicate how Theorem 1 is used in the proof.

Theorem 2. Given any three consecutive odd numbers, at least one of them is divisible by 3.

1. Any three consecutive odd numbers may be represented by $2k + 1$, $2k + 3$, and $2k + 5$.
2. If $2k + 1$ is divisible by 3, then the theorem is proved.
3. If $2k + 1$ is not divisible by 3, then either $2k + 2$ or $2k + 3$ is divisible by 3.
4. If $2k + 3$ is divisible by 3, then the theorem is proved.
5. If $2k + 3$ is not divisible by 3, then $2k + 2$ is divisible by 3.
6. If $2k + 2$ is divisible by 3, so is $2k + 5$.
7. Therefore, either $2k + 1$, $2k + 3$, or $2k + 5$ is divisible by 3.
8. Therefore at least one of any three consecutive odd numbers is divisible by 3.

By means of informal deduction, we have established a second theorem, whose truth was first suggested by observation and then proved with the aid of Theorem 1.

EXERCISE 3

Deductive Reasoning in a Logical System

Complete the following definitions by giving the class and the characteristic properties.

1. An even number is a _____ which _____.
2. An odd number is a _____ which _____.
3. A triangle is a _____ with _____.

4. A literal number is a _____ which _____.
5. A proportion is an _____ of _____.
6. A decimal fraction is a _____ whose _____.
7. A binomial is a _____ consisting of _____.
8. A decagon is a _____ with _____.
9. A median of a triangle is a _____ drawn _____.
10. A natural number is a _____ used _____.
11. A prime number is a _____ which _____.
12. A complex fraction is a _____ with _____.
13. An identity is an _____ which _____.
14. A protractor is an _____ for _____.
15. A diameter of a circle is a _____ which _____.

Criticize the following definitions.

16. An isosceles triangle is a triangle with two equal sides and two equal angles.
17. An exponent is a small number placed to the right of another number or letter.
18. Registered mail is mail which is registered.
19. An identity is a special kind of algebraic equation.
20. Parallel lines are lines which do not intersect.
21. Factoring is the process of finding the factors of a number or algebraic expression.
22. A negative number is a number which is not positive.
23. A fraction is any number less than 1.
24. An equilateral triangle is a geometric figure with equal sides.

Point out the error in each of the following syllogisms which makes the conclusions invalid.

25. Red is a warm color.
    My coat is red.
    Therefore my coat is warm.
26. Parallel lines are lines which do not intersect.
    A river and a bridge crossing it do not intersect.
    (The river and the bridge may be considered here as straight lines.)
    Therefore the river and the bridge are parallel.
27. A temperature below 32 degrees is a freezing temperature.
    A thermometer in a laboratory indicated a temperature below 32 degrees.
    Therefore it was freezing in the laboratory.
28. Cues are used in the game of billiards.
    The actor used several cues during the play.
    Therefore the actor was playing billiards.

29. A cabinet is a piece of furniture.
The President called a meeting with his cabinet.
Therefore the President called a meeting with his furniture.

30. In Problem 21 of Exercise 1, we concluded by inductive reasoning that the number formed by subtracting 1 from the square of any odd number greater than 1 is always divisible by 8. Prove this conclusion by informal deductive reasoning. *Hint.* If n is any odd number, $n + 1$ and $n - 1$ are even numbers.

31. In Problem 22 of Exercise 1, we concluded by inductive reasoning that the product of four consecutive numbers plus 1 was always a perfect square. Prove this conclusion by informal deductive reasoning. *Hint.* The four consecutive numbers may be written as k, $k + 1$, and so on.

32. In a certain textbook the Goldbach conjecture is incorrectly stated as, "Every whole number is the sum of two prime numbers." Prove by informal deduction that this statement is not true for every whole number by using the following statements:

Definition. A prime number is a number greater than 1 which has no factors other than 1 and the number itself.
Assumption. The sum of two odd numbers is always an even number.

33. Study the expression $n^2 - n + 1$ for $n = 1, 2, 3, 4$, and so forth. Is the result always an odd number? Write the given expression in the form $n(n - 1) + 1$, and try to prove your conjecture by informal deductive reasoning. *Hint.* If one of the numbers, n or $n - 1$, is odd, what about the other number? What can be said of the product $n(n - 1)$?

34. Jim Jones and his girl friend Jane Doe agree to work evenings for company X if, in addition to a bonus, the company will agree to the following terms.[7]

1. Jim is to get an evening off every ninth day.
2. Jane is to get an evening off every sixth day.
3. Jim's first day off will be on a Saturday evening.
4. Jane's first day off will be on the following Sunday evening.
5. They will be paid on their first day off together.

Do these five conditions form a logical system, that is, are they consistent and independent? Prove your conclusions by informal deductive reasoning.

4. SENTENCES, SYMBOLS, AND TRUTH TABLES

In Section 1 we described mathematics as the science of drawing conclusions from basic assumptions. The principles of deductive reasoning which

---

[7] Adapted from *Mathematician's Delight*, W. W. Sawyer, Penguin Books, Baltimore, 1951, p. 66.

we have discussed are a part of the more general science of logic. We shall make no attempt to define the term *logic*; consulting a dictionary would show that this is no easy task. We may describe logic as the systematic study of rules of inference, whether these rules apply to deducing a theorem in mathematics, solving a Sherlock Holmes mystery, or deciding on a course of action in a problem of daily living. When we say to a person, "Your arguments are not logical," we simply mean that we do not agree with his mode of thinking or reasoning and, therefore, we do not agree with his conclusions. We have already studied some of the rules of inference in an informal manner when we were concerned with the validity of a particular argument. Because validity is concerned with the form of an argument, the term "formal logic" is used to describe the rules and operations of logical reasoning. We have seen, too, the importance of words and their meanings in producing a valid argument, and the sentences that express our basic assumptions are equally important.

Sentences

There are several kinds of sentences. We are concerned with those which can be judged true or false, that is, sentences that make statements. The statement, "Washington is the capital of the United States," is true; the statement, "There are green monsters on the planet Mars," may or may not be true, but it is certainly either true or false. On the other hand, "Go to the nearest exit" is not, for purposes of logic, a statement, since we cannot judge it to be true or false. In a more subtle way, "He is captain of the football team" is not a statement, since it is not possible to judge it to be true or false until the pronoun "he" is replaced with the name of a specific person. Such sentences (that is, sentences in the ordinary sense) are called "open sentences"; the equation $x + 5 = 8$ is an open sentence until we replace x by a specific number, in which case it can be immediately judged true if x is replaced by 3, and false otherwise. Although the equation $x + 2x = 3x$ is true for every replacement of x, it is considered to be an open sentence, since no specific value of x is indicated.

The Law of Contradiction

From the foregoing discussion it is clear that we tacitly assume that no statement can be both true and false at the same time; the sentence, "It is raining," may be true at one moment and false at another but, at a given time, it is either raining or it is not raining. The assumption that a statement cannot be both true and false is one of the classical laws of logic, called the *law of contradiction*. We shall give a symbolic statement of this law later in this section.

Connectives

We shall now discuss some of the means of combining statements to make what we shall call compound sentences. These are the *connectives* "and," "or," and "not." The connective "and," whose symbol is ∧, is used to form the *conjunction* of two statements. The conjunction of the statements, "Paris is in France," and "Rome is in Italy," is written, "Paris is in France ∧ Rome is in Italy." It will be convenient to represent a statement by a letter, for example, p = "Paris is in France," and q = "Rome is in Italy." The conjunction of the two statements then becomes $p \wedge q$.

The second connective, "or," whose symbol is ∨, is used to form the *disjunction* of two statements. The disjunction of the statements p and q above is $p \vee q$, and is read, "Paris is in France or Rome is in Italy." In everyday language the word "or" has two connotations, the *inclusive* disjunction, which means one or the other or both (frequently written as and/or), and the *exclusive disjunction*, which means one or the other, but not both. The statement, "I will buy a shirt or a tie with my birthday money," may be an inclusive disjunction, since it is possible that both the tie and the shirt may be purchased. On the other hand, the statement, "I will become a lawyer or a professional dancer," is an exclusive disjunction, since it is not likely that the person will become both. We shall take the symbol ∨ to mean the inclusive disjunction, indicating by other combinations of symbols when the exclusive disjunction is intended.

The third connective, "not," whose symbol is ∼, is the denial or negation of a statement. If p = It is raining, then $\sim p$ = It is false that it is raining or, more simply, it is not raining.

Statements Expressed in Symbolic Form

We may now show how certain statements may be expressed clearly and concisely in symbolic form, using at this time only the few symbols discussed above. The first example is the law of contradiction, which takes the symbolic form $\sim(p \wedge \sim p)$. Using the parentheses in the ordinary manner, this statement says that we cannot have a statement and its denial at the same time, since if p is true, then $\sim p$ is false, which is the law of contradiction. The second example expresses the exclusive disjunction, which takes the symbolic form $(p \vee q) \wedge \sim(p \wedge q)$; that is, we have p or q and not p and q. Another symbolic form for the exclusive disjunction is $p \underline{\vee} q$.

Truth Values and Truth Tables

If a statement is true, we say that it has a truth value T, and if the statement is false, it has a truth value F. A truth table is a device for determining the truth value of a compound statement when the truth values of the components are given. We shall first show the truth tables for the basic connectives, *not*, *and*, and *or*. Table 1.1 is the truth table for the operation of negation and is, in fact, an expression of the law of contradiction. For whenever p is true, its denial is false, and when p is false, its denial is true.

Table 1.1

p	$\sim p$
T	F
F	T

Table 1.2

p	q	$p \wedge q$
T	T	T
T	F	F
F	T	F
F	F	F

Table 1.3

p	q	$p \vee q$
T	T	T
T	F	T
F	T	T
F	F	F

Tables 1.2 and 1.3 could be interpreted to represent the requirements for taking an accelerated mathematics course, where p states, "The student must have credit in Mathematics 103," and q states, "The student must have permission of the department chairman." If the connective is "and," the student may take the course only if he fulfills both requirements, which can only occur if both p and q are true. If the connective is "or," the student is disqualified only if he fails to have either of the requirements, that is, if both p and q are false. Tables 1.1, 1.2, and 1.3, may be considered as definitions of the truth values of $\sim$, $\wedge$, and $\vee$.

Obviously, many more truth tables, some of them quite complicated, can be constructed for other compound statements. We give one example here. Suppose we wish to find the truth values of the compound statement p and $\sim q$. Table 1.4 shows the various truth values for this compound statement.

Table 1.4

	p	q	$\sim q$	$p \wedge \sim q$
(1)	T	T	F	F
(2)	T	F	T	T
(3)	F	T	F	F
(4)	F	F	T	F

To illustrate Table 1.4, we may suppose the admission requirements of a college state that the candidate must be a graduate of an accredited high

26   What is Mathematical Reasoning?

school and must not have been dismissed from another college. Then if

$p = $ he must be a graduate of an accredited high school

and

$q = $ he has been dismissed from another college

the requirement is $p \wedge \sim q$. Only in case 2, where p and $\sim q$ are both true, will the requirements for admission be satisfied.

EXERCISE 4

Sentences, Symbols, and Truth Tables

Decide which of the following sentences are statements and, wherever possible, indicate their truth value. If the sentence is an open sentence, make the replacement that will make the statement true.

1. January has 31 days.
2. A certain number plus 11 equals 16.
3. All men are created equal.
4. When you come, bring a friend.
5. I hope to get an A in mathematics
6. The sum of 3^2 and 4^2 equals 7^2.
7. The product of 3^2 and $4^2 = 12^2$.
8. He was elected president for four terms.
9. Where are you going this evening?
10. All the world loves a lover.
11. $x + 5 = 11$.
12. $x + 5 = 5 + x$.
13. $x - 5 = 5 - x$.
14. Do you have the correct time?
15. The world is waiting for the sunrise.
16. All that glitters is not gold.
17. This pin is made of gold.
18. The square of a sum equals the sum of the squares.

Let A be the statement, "It is Christmas day," and B be the statement, "The bells are tolling." Write in symbolic form these compound statements.

19. It is Christmas and the bells are tolling.
20. It is Christmas or the bells are tolling.
21. It is not Christmas and the bells are tolling.
22. It is not Christmas or the bells are tolling.
23. It is not Christmas and the bells are not tolling.
24. It is not Christmas or the bells are not tolling.

Let p be the statement, "I am going to college," and q be the statement, "I will get a job." Write in words the following compound statements.

25. $p \wedge q$.
26. $p \vee q$.
27. $\sim p \wedge q$.
28. $\sim p \vee q$.
29. $\sim p \wedge \sim q$.
30. $\sim p \vee \sim q$.
31. $\sim(p \wedge q)$.
32. $(p \vee q) \wedge [\sim(p \wedge q)]$.

33. Is the compound statement of Problem 32 the inclusive disjunction or the exclusive disjunction?

Let p be the statement, "Nine is a perfect square," and q be the statement, "Fifteen is divisible by 3." Determine the truth value of each of the following compound statements. (Assume p is true and q is true.)

34. $p \vee q$.
35. $p \vee \sim q$.
36. $p \vee q$.
37. $p \wedge \sim q$.
38. $\sim p \wedge \sim q$.
39. $\sim p \vee \sim q$.

Let p be the statement, "Nine is a perfect cube," and q be the statement, "Twenty-seven is a perfect cube." Determine the truth value of each of the following compound statements. Assume p is false and q is true.

40. $p \vee q$.
41. $p \vee \sim q$.
42. $\sim p \vee q$.
43. $\sim p \vee \sim q$.
44. $p \wedge q$.
45. $p \wedge \sim q$.
46. $\sim p \wedge q$.
47. $\sim p \wedge \sim q$.

5. THE CONDITIONAL SENTENCE

Let us consider a familiar theorem of high school geometry; the angles opposite the equal sides of an isosceles triangle are equal. Frequently the student was asked to reword this theorem in order to bring out more clearly what was "given" (the hypothesis) and what was to be "proved" (the conclusion). Then this theorem was stated, "If a triangle has two equal sides, then the angles opposite these sides are equal." The theorem might also be stated, "If a triangle is isosceles, then the base angles are equal." The important point is that the connective "then" separates the hypothesis from the conclusion. In a form that takes advantage of the clarity of symbolic logic, we may restate this theorem as follows:

$p =$ A triangle has two equal sides

$q =$ The angles opposite these sides are equal

Then the statement

$$p \to q$$

which is called the *conditional*, states, "If a triangle has two equal sides, then the angles opposite these sides are equal." More briefly, the statement $p \to q$ may be read, "p implies q."[8] The statements p and q need not have a cause-and-effect relationship, as we shall see.

---

[8] Some authors use the term "implication" only for those cases where the conditional is true. We shall take the symbol $p \to q$ to mean "p implies q" whether the implication is true or false.

Truth Table for the Conditional

Since the sentences p and q in the conditional $p \to q$ may be either true or false, we shall now state the truth value of the conditional for the various truth values of p and q in Table 1.5, which is to be taken as a definition. We

Table 1.5

	p	q	$p \to q$
(1)	T	T	T
(2)	T	F	F
(3)	F	T	T
(4)	F	F	T

note that there are four cases. As an example, if p is the statement, "Roses are red," and q is the statement, "Girls are shy," then the conditional in case 4 states, "If roses are not red, then girls are not shy." Just why we assign the value T to this conditional may puzzle the student; how can we say that the conditional is either true or false? We shall give another example which we hope will make the four cases of the conditional quite plausible, although we repeat that the four cases constitute a definition and, therefore, are not to be proved.

Let p be the statement, "It is raining," and q be the statement, "The streets are wet." Then the conditional $p \to q$ states, "If it is raining, the streets are wet." In case 1 it is true that it is raining and it is also true that the streets are wet, so that we can assign the value T to the conditional $p \to q$. In case 2 it is true that it is raining but false that the streets are wet, so that the conditional $p \to q$ is obviously false. In case 3 it is false that it is raining but true that the streets are wet. This is certainly not a contradiction of the conditional that if it is raining, the streets are wet, since the streets may be wet from other causes. Therefore we assign the value T to the conditional. In case 4 it is false both that it is raining and that the streets are wet. This again is in no way a contradiction of the conditional and, again, we assign the value T to the conditional. We note that only in case 2 is there an actual contradiction of $p \to q$. It will be extremely helpful in constructing more complicated truth tables of the conditional to summarize the four cases of Table 1.5 in the following principle.

The conditional $p \to q$ is false only when p is true and q is false. In all other cases it is true.

If we call p the *antecedent* and q the *consequent*, then the principle may be stated as follows.

The conditional is false only when the antecedent is true and the consequent is false.

Variations of the Conditional

There are numerous variations of the conditional $p \to q$. We shall discuss three important variations: the *converse*, the *contrapositive*, and the *inverse*. The student may recall that in geometry, if the hypothesis and the conclusion of a theorem were interchanged, a new theorem, called the converse of the original, was formed. Thus, the theorem, "If a triangle has two equal sides then it has two equal angles," has for its converse, "If a triangle has two equal angles, then it has two equal sides." This converse happens to be true, which is not always the case for the converse. If the conditional is $p \to q$, then the converse is $q \to p$. Using the principle for constructing truth values of the conditional, Table 1.6 shows the converse. (Note that in this case, q is

Table 1.6 Converse

q	p	$q \to p$
T	T	T
T	F	F
F	T	T
F	F	T

the antecedent and p is the consequent.)

For the conditional $p \to q$, the contrapositive is $\sim q \to \sim p$, which is read, "Not q implies not p." Table 1.7 gives the truth values of the contrapositive.

Table 1.7 Contrapositive

$\sim q$	$\sim p$	$\sim q \to \sim p$
T	T	T
T	F	F
F	T	T
F	F	T

In Table 1.7 $\sim q$ is the antecedent and $\sim p$ is the consequent.

For the conditional $p \to q$, the inverse is $\sim p \to \sim q$. The truth table is given in Table 1.8.

Table 1.8 Inverse

$\sim p$	$\sim q$	$\sim p \to \sim q$
T	T	T
T	F	F
F	T	T
F	F	T

In Table 1.8 the antecedent is $\sim p$ and the consequent is $\sim q$.

Logically Equivalent Statements

To show the relationships of the four truth tables we have just discussed, we shall combine them into one table (Table 1.9).

Table 1.9

p	q	$\sim p$	$\sim q$	Conditional $p \to q$	Contrapositive $\sim q \to \sim p$	Converse $q \to p$	Inverse $\sim p \to \sim q$
T	T	F	F	T	T	T	T
T	F	F	T	F	F	T	T
F	T	T	F	T	T	F	F
F	F	T	T	T	T	T	T

A study of Table 1.9 shows that the conditional and the contrapositive have the same truth values, as do the converse and the inverse. Statements having the same truth values are said to be *logically equivalent*. The importance of logically equivalent statements is that the proof of one constitutes a proof of the other. If we prove the theorem, "The sum of two odd numbers is an even number," then we can confidently state the contrapositive, "If a number is not even, it is not the sum of two odd numbers." In the same manner, since the theorem regarding the sides and angles of an isosceles triangle has a true converse, we can state the inverse, "If two sides of a triangle are not equal, then the angles opposite these sides are not equal."

The table shows clearly another important fact; the conditional and the converse are not logically equivalent and, therefore, the truth of one does not imply the truth of the other. One may easily prove that a diagonal of a parallelogram divides it into two congruent triangles, but the converse is not true. Figure 1.8 shows two congruent triangles on either side of the diagonal

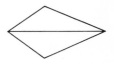

Figure 1.8

of a quadrilateral, but the quadrilateral is not a parallelogram. Thus, the conditional, "If a quadrilateral is a parallelogram, then a diagonal divides it into two congruent triangles," does not have a true converse. The converse of the conditional, "If a quadrilateral is a parallelogram, then the diagonals bisect other," does each have a true converse and, therefore, the inverse of this theorem, "If a figure is not a parallelogram, then the diagonals do not bisect each other," is also true.

THE BICONDITIONAL—NECESSARY AND SUFFICIENT CONDITIONS. We have seen that for the conditional $p \to q$, the converse $q \to p$ does not always follow. Using a nonmathematical example, let

$p =$ A man is president of the United States

$q =$ A man is a citizen of the United States

Then $p \to q$ asserts that if a man is president of the United States, he is a citizen of the United States. An alternate way of stating this conditional is that a *necessary* condition for a man to be President of the United States is that he is a citizen of the United States; he cannot be president if he is not a citizen. We note that the converse is certainly not true; if a man is a citizen it does not follow that he is the president. We state this fact by saying that being a citizen of the United States is not a *sufficient* condition for being the president.

A different situation arises when we let

$p =$ A man has committed a crime

$q =$ A man cannot hold public office

Then $p \to q$ asserts that if a man has committed a crime, he cannot hold public office. In this case the commission of a crime is sufficient to bar a man from public office. Here, again, the converse $q \to p$ is certainly not true; a man who is barred from public office may not have been guilty of a crime. The commission of a crime is not a necessary condition for being denied public office.

We now illustrate the case where both the conditional $p \to q$ and the converse $q \to p$ are true. We state this as

$$(p \to q) \land (q \to p)$$

which is usually abbreviated as

$$p \leftrightarrow q$$

Either of these statements expresses the *biconditional*. To illustrate the biconditional let

$p =$ A number is divisible by 3

$q =$ The sum of the digits of a number is divisible by 3

32   *What is Mathematical Reasoning?*

It is shown in arithmetic textbooks that $p \leftrightarrow q$, that is, if a number is divisible by 3, the sum of its digits is divisible by 3, and if the sum of the digits of a number is divisible by 3, the number is divisible by 3. In terms of necessary and sufficient conditions, we may state the biconditional as follows.

The necessary and sufficient condition that a number is divisible by 3 is that the sum of its digits is divisible by 3.

By way of comparison, the statements

$$p = \text{A number is divisible by 4}$$
$$q = \text{A number is divisible by 2}$$

do not give a true biconditional since, although $p \to q$, it does not follow that $q \to p$ because 18 is divisible by 2 but not by 4. Thus, divisibility of a number by 2 is a necessary but not sufficient condition for divisibility by 4.

Still another way of stating the biconditional is suggested by the truth table for the biconditional shown in Table 1.10. Cases 1 and 4 are the only cases

Table 1.10

	p	q	$p \to q$	$q \to p$	$p \leftrightarrow q$
(1)	T	T	T	T	T
(2)	T	F	F	T	F
(3)	F	T	T	F	F
(4)	F	F	T	T	T

where the biconditional is true. In case 1 both p and q are true and, in case 4, both p and q are false. We may then say that, for a true biconditional, q is true if and only if p is true. For the example above on divisibility by 3, we may state that a number is divisible by 3 if and only if the sum of its digits is divisible by 3. In summary, for this particular biconditional the following statements are equivalent.

1. $p \leftrightarrow q$.
2. The necessary and sufficient condition that a number is divisible by 3 is that the sum of its digits is divisible by 3.
3. A number is divisible by 3 if and only if the sum of its digits is divisible by 3.

We state one more form of the conditional, using the connective *only if*. We illustrate the meaning of this connective with the statement, "I will learn to drive only if I buy a car." Ordinarily the *if* designates the antecedent, but we see that the phrase *only if* in the above conditional implies that if

I learned to drive I must have bought a car. Therefore the statement, "*p* only if *q*" is symbolically expressed as $p \to q$.

EXERCISE 5

THE CONDITIONAL SENTENCE

Let A = It is not cold
B = It is not raining
C = We will go on a picnic

State the following sentences in symbolic form.

1. If it is not raining we will go on a picnic.
2. If it rains we will not go on a picnic.
3. If it is cold we will not go on a picnic.
4. If it is not cold we will go on a picnic.
5. We will go on a picnic if and only if it is not cold.
6. We will go on a picnic if and only if it is not raining.
7. We will go on a picnic if it is not cold and it is not raining.
8. If it is not cold or it is not raining we will go on a picnic.
9. We will go on a picnic if and only if it is not raining and it is not cold.
10. We will not go on a picnic if and only if it is cold.
11. We will not go on a picnic if and only if it is not raining.
12. We will not go on a picnic if and only if it is not cold and it is not raining.

Let p = I brush my teeth (T)
q = I drink a quart of milk daily (F)
r = I avoid excessive sweets (T)
s = I will have no cavities in my teeth (F)

Find the truth value of each of the following statements.

13. $(p \wedge q) \to s$.
14. $(p \vee q) \to s$.
15. $(p \wedge r) \to s$.
16. $(p \vee r) \to s$.
17. $s \to (p \wedge r)$.
18. $s \to (p \wedge q)$.
19. $s \to (p \vee r)$.
20. $(p \wedge \sim r) \to \sim s$.
21. $(p \wedge \sim q) \wedge r \to \sim s$.
22. $(p \wedge \sim q) \wedge r \to s$.
23. $(-p \wedge \sim r) \wedge q \to \sim s$.

State in words the converse, inverse, and contrapositive of each of the following conditionals.

24. If two numbers are odd their sum is even.
25. If two numbers are even their sum is even.
26. If a first number is odd and a second number is even, their product is even.
27. If two numbers are odd their product is odd.
28. If two numbers are even their product is even.

29. If a number is a perfect square it has two equal factors.
30. If a number is a perfect cube it has three equal factors.
31. The converse of the following theorem is known to be true. One root of the quadratic equation $ax^2 + bx + c = 0$ is equal to zero if $c = 0$. State the theorem in terms of necessary and sufficient conditions.

In Problems 32 to 40, let $p = $ I will work this summer, and $q = $ I will get a car, and express the given conditionals in symbolic form.

32. If I work this summer I will get a car.
33. If I get a car I will work this summer.
34. I will get a car only if I work this summer.
35. I will get a car if I work this summer.
36. I will get a car if and only if I work this summer.
37. If I do not work this summer I will not get a car.
38. I will not get a car if and only if I do not work this summer.
39. If I get a car it is necessary and sufficient that I work this summer.
40. If I do not get a car I will not work this summer.

6. VALID ARGUMENTS AND TAUTOLOGIES

We are now in a position to return to the central idea of this chapter: the validity of an argument. We shall define a valid argument as an argument for which the conclusion is true whenever the assumptions are true. Let us illustrate this definition by the following argument: If John feels ill he will not go to the party. John will go to the party. Therefore John does not feel ill. Let

$$p = \text{John feels ill}$$
$$q = \text{John will go to the party}$$

Then the assumptions are

$$p \rightarrow -q \quad \text{(If John feels ill he will not go to the party)}$$

and
$$q \quad \text{(John will go to the party)}$$

and the conclusion is

$$-p \quad \text{(John does not feel ill)}$$

To prove that the argument is valid, we must show that the conclusion, $-p$, is true when the assumptions, $p \rightarrow -q$ and q, are true.

6. Valid Arguments and Tautologies

The first three columns of Table 1.11 are written to determine the cases for

Table 1.11

	p	$-q$	$p \rightarrow -q$	q	$-p$
(1)	T	T	T	F	F
(2)	T	F	F	T	F
(3)	F	T	T	F	T
(4)	F	F	T	T	T

which the assumption $p \rightarrow -q$ is true. The fourth column is written to show all cases for which the assumption q is true. The last column, the negation of the first column, indicates when the conclusion $-p$ is true. Now case 4 is the only case where both assumptions are true, and for this case the conclusion is true. Hence the argument is valid by definition. Had there been any case where the assumptions were true and the conclusion not true, the argument would not be valid.

As a second example, let us test the validity of the following argument: If Marie is French, she is beautiful. Marie is not French. Therefore she is not beautiful. Let

$$p = \text{Marie is French}$$
$$q = \text{Marie is beautiful}$$

Then the assumptions are

$$p \rightarrow q$$

and

$$\sim p$$

and the conclusion is

$$\sim q$$

The truth table is constructed in Table 1.12. In case 3 the assumptions are

Table 1.12

	p	q	$p \rightarrow q$	$\sim p$	$\sim q$
(1)	T	T	T	F	F
(2)	T	F	F	F	T
(3)	F	T	T	T	F
(4)	F	F	T	T	T

true, but the conclusion is false. Therefore, the argument is not valid. Note that in a simple argument such as this we can recognize that the argument is

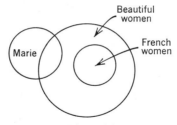

Figure 1.9

not valid without constructing the truth table, by drawing a diagram similar to those in Section 2. The assumption $p \to q$ is equivalent to the statement, "All French women are beautiful." Then, in Figure 1.9, the outer circle represents beautiful women and the inner circle represents French women. The assumption, "Marie is not French," merely places her outside the circle of French women but does not exclude her from the outer circle of beautiful women: there is, therefore, no conclusion. Stated in another way, we see that the assumption $p \to q$ is a sufficient but not necessary condition for being beautiful.

Tautologies

The first example in this section may be stated as follows:

$$[(p \to \sim q) \land q] \to \sim p$$

This may be read as follows: "If p implies that q is false, and q is true, then p is false." Let us construct a truth table for this conditional (Table 1.13).

Table 1.13

p	q	$\sim q$	$p \to \sim q$	$(p \to \sim q) \land q$	$\sim p$	$[(p \to \sim q) \land q] \to \sim p$
T	T	F	F	F	F	T
T	F	T	T	F	F	T
F	T	F	T	T	T	T
F	F	T	T	F	T	T

What is significantly different in this table from the other truth tables we have constructed for the conditional is that this conditional, $[(p \to \sim q) \land q] \to \sim p$ is true for all values of p and q. A compound statement that is true for all values of p and q is called a *tautology*. It is possible to show that a tautology represents a valid argument and is, in fact, a consequence of the definition of a valid argument given earlier in this section.[9]

The Syllogism as a Valid Argument

We can now prove that the syllogism, discussed in Section 2, is a valid argument. Consider the syllogism of Aristotle:

> All men are mortal.
> Socrates is a man.
> Therefore Socrates is mortal.

Let
p = He is a man.
q = He is mortal
r = He is Socrates

we restate the syllogism as
If he is a man, then he is mortal $(p \to q)$.
If he is Socrates, then he is a man $(r \to p)$.
Therefore, if he is Socrates, then he is mortal $(r \to q)$.

In symbolic form, this becomes the Law of Syllogism,[10]

$$[(p \to q) \land (r \to p)] \to (r \to q)$$

We have only to show that this is a tautology to prove that the syllogism is a valid argument. We show this in Table 1.14.

Table 1.14

p	q	r	$p \to q$	$r \to p$	$(p \to q) \land (r \to p)$	$r \to q$	$[(p \to q) \land (r \to p)]$ $\to (r \to q)$
T	T	T	T	T	T	T	T
T	F	T	F	T	F	F	T
T	T	F	T	T	T	T	T
T	F	F	F	T	F	T	T
F	T	T	T	F	F	T	T
F	T	F	T	T	T	T	T
F	F	T	T	F	F	F	T
F	F	F	T	T	T	T	T

[9] For a proof of this, see *Symbolic Logic and the Real Number System*, A. H. Lightstone, Harper and Row, New York, 1965, p. 21.
[10] The Law of Syllogism is usually written $[(p \to q) \land (q \to r)] \to (p \to r)$ which is equivalent to our statement.

The last column consists of all *T*'s. The argument is a tautology and, therefore, valid. Although we used a specific syllogism for purposes of illustration, the proof applies to all syllogisms.

EXERCISE 6

Valid Arguments and Tautologies

Test the validity of the following arguments by the definition given in the first paragraph of this section, or by the use of diagrams.

1. Students who excel in music are good in mathematics. This student excels in music. Therefore he is good in mathematics.
2. Students who excel in music are good in mathematics. This student does not excel in music. Therefore he is not good in mathematics.
3. Students who excel in music are good in mathematics. This student is good in mathematics. Therefore he excels in music.
4. Students who excel in music are good in mathematics. This student is not good in mathematics. Therefore he does not excel in music.
5. Only students who excel in music are good in mathematics. This student excels in music. Therefore he is good in mathematics.
6. Only students who excel in music are good in mathematics. This student does not excel in music. Therefore he is not good in mathematics.
7. Only students who excel in music are good in mathematics. This student is good in mathematics. Therefore he excels in music.
8. Only students who excel in music are good in mathematics. This student is not good in mathematics. Therefore he does not excel in music.

Test the validity of the following arguments by showing that they are or are not tautologies.

9. When Mary expects to be late, she calls home. Mary expects to be late. Therefore she called home.
10. When Mary expects to be late, she calls home. Mary will call home. Therefore she expected to be late.
11. When Mary expects to be late, she calls home. Mary did not call home. Therefore she did not expect to be late.
12. When Mary expects to be late, she calls home. Mary does not expect to be late. Therefore she will not call home.
13. Mary calls home only when she expects to be late. Mary called home. Therefore she expected to be late.
14. Mary calls home only when she expects to be late. Mary did not call home. Therefore she did not expect to be late.

15. Mary calls home only when she expects to be late. Mary expected to be late. Therefore she called home.
16. Mary calls home only when she expects to be late. Mary did not expect to be late. Therefore she did not call home.
17. Is $p \rightarrow (p \wedge q)$ a tautology?
18. Is $p \rightarrow (p \vee q)$ a tautology?
19. Is $(p \wedge q) \rightarrow p$ a tautology?
20. Is $(p \vee q) \rightarrow p$ a tautology?
21. The Law of the Excluded Middle states that either p is true or p is false. Symbolically, this is $p \vee \sim p$. Show that this is a valid law by proving it is a tautology.
22. The Law of Contradiction, discussed in Section 4, may be expressed as $\sim(p \wedge \sim p)$. Show that this is a tautology.
23. Test the validity of the following argument: If ignorance is bliss it is folly to be wise. It is not foolish to be wise. Therefore ignorance is not bliss.
24. Test the validity of the following argument: If John does not do his homework he will fail the course. John failed the course. Therefore John did not do his homework.
25. Consider the following argument.

    If it is a good book it is worth reading.
    It is worth reading only if the plot is simple.
    The plot is not simple.
    Therefore it is not a good book.

    If we have

    $p =$ It is a good book
    $q =$ It is worth reading
    $r =$ The plot is simple

    (a) Write the argument in symbolic form.
    (b) Show that the argument is valid.

7. THE CONCEPT OF SET

Before we begin the study of sets, we may remark that we have worked with sets in previous sections without specifically referring to them as sets. The "even numbers," the "beautiful women," the "case histories of diabetics," and so forth, are collections of objects (physical or mental) which, in mathematics, are referred to as "sets." We shall find that the subjects of sets and logic are closely related, to each other and to the entire structure of mathematics.

"The theory of sets stands as one of the boldest and most beautiful creations of the human mind; its construction of concept and method of proof have reanimated and revitalized all branches of mathematical study." So states Joseph Breuer in the introduction to *Theory of Sets*.[11] Since a set may be described as a "well defined collection of objects" (we shall elaborate on this description later), we are naturally led to inquire how a concept as simple as this can be so basic to the study of mathematics. We shall try to show this in what follows.

We begin our inquiry into the nature of sets by considering the ninth axiom of Euclid; the whole is greater than any of its parts. In Figure 1.10, segment $\overline{PB}$ is a part of segment $\overline{AB}$, and we should therefore expect, by Euclid's ninth axiom, that $\overline{AB}$ is greater than $\overline{PB}$. As a matter of fact, even without benefit of Euclid's axiom, we would not hesitate to say that this is true.

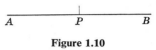

Figure 1.10

In Figure 1.11, Segment $\overline{P'B'}$ is drawn parallel and equal to segment $\overline{PB}$, and $\overline{AP'}$ and $\overline{BB'}$ are drawn so as to intersect in point Q. We should expect, of course, that $\overline{AB}$ is greater than $\overline{P'B'}$. If now, as in Figure 1.12, lines are drawn from Q intersecting segment AB in points a, b, c, etc., and intersecting segment $\overline{P'B'}$ in points a', b', c', etc., then for every point a, there is a point a', and for every point a', there is a point a. It is clear that there are as many points on a part of the segment, that is, on $\overline{P'B'}$, as there are on the whole segment $\overline{AB}$. Since a point is dimensionless, we cannot attribute this startling fact to any difference in the size of the points and, if we consider the totality of points on a line segment as equivalent to the segment itself, there is at once evident a contradiction of the ninth axiom of Euclid (as well as a contradiction of our own intuition and experience); the part is equal to the whole.

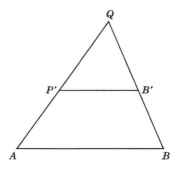

Figure 1.11

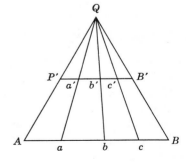

Figure 1.12

[11] *Introduction to the Theory of Sets*, Joseph Breuer, translated by Howard Fehr, Prentice Hall, Inc., 1959.

7. The Concept of Set

We can illustrate a similar situation by a pairing of the odd and even numbers with the even numbers alone:

$$1 \quad 2 \quad 3 \quad 4 \quad 5 \quad 6 \quad 7 \quad 8 \quad 9 \quad 10 \cdots$$
$$2 \quad 4 \quad 6 \quad 8 \quad 10 \quad 12 \quad 14 \quad 16 \quad 18 \quad 20 \cdots$$

where the three dots indicate that this may be continued as far as we please. We would expect that there would be fewer even numbers than odd and even numbers together, since the even numbers are but part of the whole array of odd and even numbers. But since there is no end to the pairing of these two rows of numbers, we can only conclude that there are as many even numbers as odd and even numbers together, and once again, the part is equal to the whole.

It is interesting to note that Galileo, as early as 1638, observed, by a similar pairing of numbers and their squares, that there were as many squares of integers as there were integers.[12] Through the words of his character, Salviatus, he says, "Therefore if I assert that all numbers, including both squares and non-squares, are more than the squares alone, I shall speak the truth, shall I not?" After a long discussion, he continues, "So far as I see we can only infer that the totality of all numbers is infinite, that the number of squares is infinite, and that the number of their roots is infinite; neither is the number of squares less than the totality of all numbers . . . ." In fact, in the words of Bertrand Russell,[13] "It is this property (the equivalence of a part of the numbers with the whole) which was formerly thought to be a contradiction, that is now transformed into a harmless definition of infinity. . . ."

If, as we now suspect, the contradictions of Euclid's ninth axiom are related to the behavior of infinite "sets" of numbers or infinite "sets" of points on a line segment, we can get some inkling of the importance of the notion of sets. In fact, the theory of sets was created by Georg Cantor (1845–1918) to establish a theory of infinity in a mathematically acceptable way. But, as is often the case in scientific research, the results of Cantor's work went far beyond the original purpose, and it has been said that the mathematical concept of a set can be used as the foundation for all known mathematics.[14] In our introduction to the subject of sets we shall confine ourselves chiefly to finite sets, taking the term *finite* to mean, for the time being, that the sets will contain a definite number of objects, or *elements*, as we shall call them. The 26 letters of the alphabet, or the first 100 even numbers, are examples of finite sets, as contrasted to the set of points on a line, or the set of all numbers, which are infinite sets.

---

[12] *Discorsie Dimonstrazione Matematiche Intorno a Due Nuove Scienze*, Leyden, 1638.

[13] "Mathematics and Metaphysicians," Bertrand Russell, *The World of Mathematics*, Volume 3, Simon and Schuster, 1956.

[14] "Reproduced from *Naive Set Theory*, by *P. Halmos*, by permission of Van Nostrand-Reinhold Company, a division of Litton Educational Publishing, Inc., Litton Industries, *Princeton, New Jersey, 1960.*

The Meaning of Set

Cantor conceived a set as a collection of definite, well-distinguished objects of our perception or of our thought. It is clear that numbers, which will frequently be elements of the sets we will discuss, exist in the mind rather than as physical objects; that theoretically, at least, it must be possible to determine whether or not a particular element belongs to a given set; and as a consequence of the term "well-distinguished," that no element may appear more than once in a given set. It is quite easy to determine, for example, that the number 7 is an element of the set of prime numbers and that the number 22 is not an element of that set. On the other hand, it might be quite difficult to determine whether or not the number $2^{127} - 1$ is an element of the set of prime numbers, but the answer is difficult rather than debatable. The prime factors of 12, listed as 2, 2, 3, do not form a set, since the element 2 is repeated. It is quite important to note that we have not defined the word *set*, we have simply used one of its synonyms, *collection*, to describe it.

Designating a Set

A set may be specified by naming the members or elements; the collection

$$\{book, spoon, coat\}$$

is such a set, the braces indicating set membership. In more general use is the method of specifying a set by stating a common property or properties, which the elements are to possess. For example, the set of positive even numbers less than 10 is $\{2, 4, 6, 8\}$, and the set of vowels of the English alphabet is $\{a, e, i, o, u\}$. We frequently name the set by a capital letter, so that if A is the set of vowels of the English alphabet, we have

$$A = \{a, e, i, o, u\}$$

To indicate that the letter a is a member of set A, we use the symbol $\in$, and we write

$$a \in A$$

Since the letter j is not a member of set A, we may write

$$j \notin A$$

We may also designate the set A in what is termed "set builder" notation; thus

$$A = \{x \mid x \text{ is a vowel of the English alphabet}\}$$

The vertical bar is to be read "such that," and the whole expression reads: "A is the set of all elements x, such that x is a vowel of the English alphabet."

Similarly, the expression

$$B = \{x \mid x \text{ is a whole number, and } 1 < x < 10\}$$

is read: "B is the set of all elements x, such that x is a whole number between 1 and 10." The elements of this set are 2, 3, 4, 5, 6, 7, 8, 9, and we may also write

$$B = \{2, 3, 4, 5, 6, 7, 8, 9\}$$

Classification of Sets

We have discussed, but not defined, infinite sets and finite sets. There are several other ways in which sets may be classified. A *unit* set has one element, an example being the set of months with only 28 days. The only element is February. An *empty* set is a set with no elements, an example being the set of months with 32 days. The empty set is also called the *null* set, and its symbol is $\emptyset$ or { }.

Frequently we may wish to confine ourselves to working with only part of the elements of a given set, as, for example, the vowels in the set of letters of the alphabet. Such a set is called a *subset*, and the entire set is called the *universal* set, which is usually designated by the symbol U. If we designate the set of vowels by the letter V, then we write

$$V \subset U$$

which reads, "V is a subset of U." It is convenient to have a precisely stated definition of a subset:

Definition 1. Set A is a subset of set B if and only if every element of A is an element of B.

In Section 8 we shall give another definition of a subset and a diagram that illustrates both definitions.

Number of Subsets from the Universal Set

Suppose the universal set U is the set $\{a, b, c\}$. The question naturally arises, "How many subsets can be constructed from the set U?" Let us list these subsets; $\{a, b, c\}$; $\{a, b\}$; $\{a, c\}$; $\{b, c\}$; $\{a\}$; $\{b\}$; $\{c\}$; { }. There are a number of observations to be made concerning these eight subsets. First, we note that the set U itself, that is, the set $\{a, b, c\}$ is considered a subset of itself. Next we observe that the empty set is also one of the subsets. Why is this so? From the definition of a subset it follows that any subset A is not a subset of a universal set U only if there is at least one element in A that is not an element of U. But the empty set contains no elements and hence cannot contain any element that is not an element of U. Therefore we must include the empty set among the subsets of U. Finally, we observe that there are exactly eight subsets. The set of all subsets of the universal set U is called the *power set* of U. The name power set stems from the fact that the number of subsets of a universal set of n elements is equal to 2^n, that is, the nth power of 2. Some students may recognize that this formula is the formula for the total number of combinations of n things taken 1, 2, 3, 4,—or n at a time. One more remark is in order. The term *proper subset* is reserved for those subsets which do not contain all of the elements of the universal set. In the example above, the set $\{a, b, c\}$ is not a proper subset of the set $U = \{a, b, c\}$, but all of the other subsets are. Some authors use the symbol $A \subset B$ for proper subsets only, using the symbol $A \subseteq B$ for any subset.

The Complement of a Set

If set A is a subset of the universal set U, then there are elements of U that are not elements of A. We denote the set of these elements by the symbol $\bar{A}$, which is called the complement of A. If the set U is the set $\{1, 2, 3, 4, 5, 6\}$, and the subset A is the set $\{2, 4, 6\}$, then $\bar{A}$ is the set $\{1, 3, 5\}$.

We may state the definition of the complement of a set in symbolic form as follows.

Definition 2.
$$\bar{A} = \{x \mid x \in U \text{ and } x \notin A\}$$

Figure 1.13 shows the relation of U, A, and $\bar{A}$ for the sets just discussed.

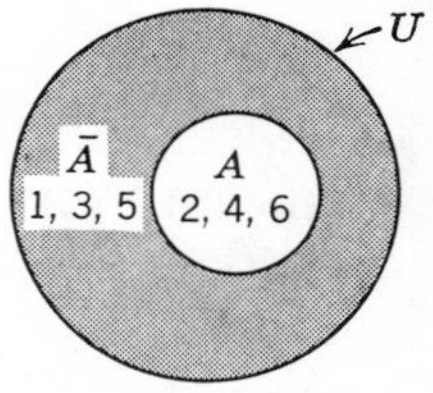

Let us now summarize the various aspects of a set. Let the universal set U be the set

$$U = \{x \mid x \text{ is divisible by 2 and } 1 < x < 11\}$$

Figure 1.13

1. $U = \{2, 4, 6, 8, 10\}$
2. A proper subset of U is the set A, where
$$A = \{x \mid x \text{ is divisible by 4, and } x \in U\}$$
$$A = \{4, 8\}$$
3. A proper subset of U, which is a unit set, is the set B, where
$$B = \{x \mid x \text{ is divisible by 8, and } x \in U\}$$
$$B = \{8\}$$
4. One specification for the proper subset of U, which is the empty set, is the set C, where
$$C = \{x \mid x \text{ is divisible by 7 and } x \in U\}$$
$$C = \{\ \}$$

5. The complement of set A is the set $\bar{A}$, where

$$\bar{A} = \{2, 6, 10\}$$

6. The power set of U has 2^5, or 32, elements, since U has 5 elements.

EXERCISE 7

THE CONCEPT OF SET

1. Let U represent the set of all positive integers less than or equal to 10. List the elements of the following sets, using set notation.

   (a) The set U.
   (b) The subset A whose elements are divisible by 2.
   (c) The subset $\bar{A}$.
   (d) The subset B whose elements are solutions of $x^2 - 7x + 10 = 0$.

2. How many sets are in the power set of U?
3. If $U = \{0, 1, 2, 3, 4, 5\}$, list the elements of the following subsets.

   (a) $\{x \mid x \in U \text{ and } x > 0\}$
   (b) $\{x \mid x \in U \text{ and } x < 4\}$
   (c) $\{x \mid x \in U \text{ and } 2 < x < 5\}$
   (d) $\{x \mid x \in U \text{ and } x \neq 5\}$
   (e) $\{x \mid x \in U \text{ and } x + 7 = 11\}$
   (f) $\{x \mid x \in U \text{ and } x + 7 = 13\}$

4. If $U = \{2, 4, 6, 8, 10\}$, list the elements of the following subsets.

   (a) $\{x \mid x \in U \text{ and } x + 2 \in U\}$
   (b) $\{x \mid x \in U \text{ and } x + 3 \in U\}$
   (c) $\{x \mid x \in U \text{ and } \dfrac{x}{2} \in U\}$
   (d) $\{x \mid x \in U \text{ and } 2x \in U\}$
   (e) $\{x \mid x \in U \text{ and } x^2 \in U\}$

5. Write a description of each of the following sets.

   (a) $A = \{2, 4, 6, 8, 10\}$
   (b) $B = \{1, 3, 5, 7, 9\}$
   (c) $C = \{1, 4, 9, 16, 25\}$
   (d) $D = \{1, 8, 27, 64, 125\}$
   (e) $E = \{4, 14, 24, 34, 44\}$
   (f) $F = \{1/2, 1/3, 1/4, 1/5, 1/6, 1/7, 1/8, 1/9\}$

6. Write the set of integers n such that $n^2 + 2n + 1$ is not a perfect square.
7. List 5 elements of set A, where $A = \{x \mid x > 2x\}$.
8. List 5 elements of set B, where $B = \{x \mid x^2 < x\}$.

46   What is Mathematical Reasoning?

9. Let $U = \{a, b, c, d, e, f, g\}$, $A = \{a, c, f\}$, and $B = \{a, b, f, g\}$. List the elements of the set $\bar{A}$ and the set $\bar{B}$.
10. The complement of the complement of set A is written $\bar{\bar{A}}$. For sets A and B of Problem 9, find $\bar{\bar{A}}$ and $\bar{\bar{B}}$. Can you make a general statement concerning the set $\bar{\bar{A}}$?
11. What is the complement of the universal set U?
12. What is the complement of the empty set?

The following statements apply to Problems 13–16.

A is the set of triangles having two or more equal sides.
B is the set of equilateral triangles.
C is the set of squares.
D is the set of rectangles.

Recalling that set S is not a subset of set U if there is at least one element of S which is not an element of U, state whether the following statements are true or false.

13. $A \subset B$.   14. $B \subset A$.   15. $C \subset D$.   16. $D \subset C$.
17. Describe set A where $A = \{x \mid x \neq x\}$.

If $A = \{x \mid x = 1, 2, 3, 4, 5, 6, 7, 8, 9\}$ and $B = \{y \mid y = -1, -2, -3, -4, -5, -6, -7, -8, -9\}$:

18. Is $x + y$ ever an element of set A? Illustrate your answer with an example.
19. Is $x + y$ ever an element of set B? Illustrate your answer with an example.
20. Is xy ever an element of set A? Illustrate your answer with an example.
21. Is xy ever an element of set B? Illustrate your answer with an example.
22. Is y^2 ever an element of set A? Illustrate your answer with an example.
23. Is y^2 ever an element of set B? Illustrate your answer with an example.
24. Is x^2 ever an element of set B? Illustrate your answer with an example.

8. RELATED SETS

Just as we have certain relations between numbers, such as "equal to," "greater than," and so on, and certain operations on them, such as addition and subtraction, we have relations between sets and operations on them. We shall discuss some of these relations and operations.

Equality of Sets

The first relation we consider is that of equality, for which we have the following definition:

Definition 3. Two sets, A and B, are equal if every element of A is an element of B, and every element of B is an element of A.

This definition may be stated symbolically as follows:

$$A = B \text{ if and only if } A \subset B \text{ and } B \subset A$$

The sets $\{a, b, c\}$, $\{b, c, a\}$, and $\{a, c, b\}$ are equal sets. We note in particular that the equality of sets does not merely refer to the number of elements in each set, although it is implicit in the definition that the number of elements in each of the equal sets is the same. The sets $\{a, b, c\}$ and $\{a, b, d\}$ are not equal because their elements are not the same, although they each have three elements.

Union of Sets

The first operation we consider is the *union* of two sets, for which we state the following definition.

Definition 4. The union of two sets, A and B, is the set of all elements which are either in A or in B or in both A and B.

The symbol for union is $\cup$, and the expression $A \cup B$ means the union of A and B. If $A = \{a, b, c\}$ and $B = \{b, c, d\}$, then $A \cup B$ is the set $\{a, b, c, d\}$. It is important to understand that the elements that are in both sets are not repeated; b and c are elements of both sets but are included only once in the union. A symbolic definition of the union of two sets may be written

$$A \cup B = \{x \mid x \in A \text{ or } x \in B\}$$

Diagrams for the Union of Sets

The union of sets, as well as other set operations, can be illustrated in much the same manner as the deductive arguments were illustrated in Section 2. We wish to emphasize that the diagrams serve only as an aid to visualizing set relations and operations but do not prove the results. We give three examples of the union of two sets and their diagrams. The shaded part of each diagram shows the part of each set included in the union.

EXAMPLE 1 (Figure 1.14). $A = \{1, 2, 3\}$. $B = \{2, 3, 4\}$. $A \cup B = \{1, 2, 3, 4\}$.

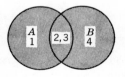

Figure 1.14

48   *What is Mathematical Reasoning?*

EXAMPLE 2 (Figure 1.15).

$A = \{1, 2, 3\}$. $B = \{4, 5, 6\}$. $A \cup B = \{1, 2, 3, 4, 5, 6\}$.

Figure 1.15

EXAMPLE 3 (Figure 1.16).

$A = \{1, 2, 3, 4, 5, 6\}$. $B = \{2, 3, 4\}$. $A \cup B = \{1, 2, 3, 4, 5, 6\}$.

Figure 1.16

Intersection of Sets

The second operation to be defined is the *intersection* of two sets.

Definition 5. The intersection of two sets, A and B, is the set of all elements that are elements of both A and B.

The symbol for intersection is $\cap$, and the expression $A \cap B$ means the intersection of A and B. If $A = \{a, b, c\}$, and $B = \{b, c, d\}$, then $A \cap B$ is the set $\{b, c\}$. A symbolic definition of the intersection of two sets may be written

$$A \cap B = \{x \mid x \in A \text{ and } x \in B\}$$

It may be helpful to observe that the operation of the intersection of two sets is analogous to the idea of intersection in geometry, since if two straight lines intersect, the point of intersection is that point which is on both lines.

Disjoint Sets

It may happen that $A \cap B$ is the set $\varnothing$, that is, the empty set. This means that the two sets have no elements in common. In this case, A and B are called *disjoint* sets. The sets $\{a, b, c\}$ and $\{d, e, f, g\}$ are disjoint sets.

Diagrams for the Intersection of Sets

We give three examples of the intersection of two sets and their diagrams.

EXAMPLE 4 (Figure 1.17). $A = \{1, 2, 3\}$. $B = \{2, 3, 4\}$. $A \cap B = \{2, 3\}$.

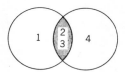

Figure 1.17

EXAMPLE 5 (Figure 1.18). $A = \{1, 2, 3\}$. $B = \{4, 5, 6\}$. $A \cap B = \{\ \}$. A and B are disjoint.

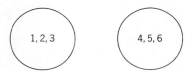

Figure 1.18

EXAMPLE 6 (Figure 1.19). $A = \{1, 2, 3, 4, 5, 6\}$. $B = \{2, 3, 4\}$. $A \cap B = \{2, 3, 4\}$.

Figure 1.19

The shaded part of each diagram shows the part of each set included in the intersection.

Subsets Defined in Terms of Union and Intersection of Sets

The operations of union and intersection of sets make it possible to state the definition of a subset in concise symbolic form as follows.

Definition 6:
$$A \subset B \leftrightarrow A \cup B = B$$

Definition 7:
$$A \subset B \leftrightarrow A \cap B = A$$

Figures 1.20 and 1.21 illustrate these definitions.

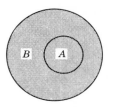
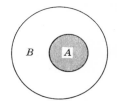

Figure 1.20   Figure 1.21

De Morgan's Laws

The operations of union and intersection of sets give many interesting relations. We illustrate two of these, which are known as De Morgan's Laws. The first of these laws is

$$\overline{A \cup B} = \bar{A} \cap \bar{B}$$

To illustrate (but not prove) this relation, let

$$U = \{1, 2, 3, 4, 5, 6, 7, 8\}$$
$$A = \{1, 3, 5\}$$
$$B = \{1, 4, 6\}$$

Then we may write

$$\bar{A} = \{2, 4, 6, 7, 8\}$$
$$\bar{B} = \{2, 3, 5, 7, 8\}$$

Then we shall have

$$A \cup B = \{1, 3, 4, 5, 6\}$$

and
$$\overline{A \cup B} = \{2, 7, 8\} \tag{1}$$
We also have
$$\bar{A} \cap \bar{B} = \{2, 7, 8\} \tag{2}$$

Since the sets in 1 and 2 are equal, we have verified that $\overline{A \cup B} = \bar{A} \cap \bar{B}$ for the given sets A and B. The second De Morgan Law is

$$\overline{A \cap B} = \bar{A} \cup \bar{B}$$

We have
$$A \cap B = \{1\}$$

and
$$\overline{A \cap B} = \{2, 3, 4, 5, 6, 7, 8\} \tag{3}$$
We also have
$$\bar{A} \cup \bar{B} = \{2, 3, 4, 5, 6, 7, 8\} \tag{4}$$

Since the sets in 3 and 4 are equal, we have verified that $\overline{A \cap B} = \bar{A} \cup \bar{B}$ for the given sets A and B.

RELATION BETWEEN CONNECTIVES AND SET OPERATIONS. It is interesting to note that the union of two sets plays a role analogous to the disjunction of two statements and that the intersection of two sets is analogous to the conjunction of two statements. Consider the disjunction, "They are either Masons or Elks." The word "they" refers to people who are either Masons or Elks, or both. The symbolic statement is then $M \vee E$. But since M represents a set of Masons and E represents a set of Elks, we may also write, in set notation, $M \cup E$, that is, the union of M and E. It appears that the logical connective "or" corresponds to the set operation of union.

If we now consider the conjunction, "They are Masons and Elks," the word "they" refers to people who are both Masons and Elks. The symbolic statement is $M \wedge E$. But the intersection of two sets comprises elements which belong to both sets; therefore, the set notation is $M \cap E$. It appears that the logical connective "and" corresponds to the set operation of intersection.

Finally, consider the negation of M, which is $\sim M$. This refers to people who are not Masons. But $\bar{M}$ is the complement of M and, therefore, the negation of a statement corresponds to the complement of a set.

De Morgan's Laws furnish us with an excellent example of the correspondence of the symbols of logic and the symbols of set operations. Thus, the first De Morgan Law,

$$\overline{A \cup B} = \bar{A} \cap \bar{B}$$

expressed in symbolic logic is

$$\sim(A \vee B) \leftrightarrow \sim A \wedge \sim B$$

The second De Morgan Law,

$$\overline{A \cap B} = \bar{A} \cup \bar{B}$$

expressed in symbolic logic is

$$\sim(A \wedge B) \leftrightarrow \sim A \vee \sim B$$

EXERCISE 8

RELATED SETS

Given $U = \{0, 1, 2, 3, 4, 5\}$, $A = \{0, 1, 2, 3\}$, and $B = \{1, 2, 3, 4\}$, find the following sets. Illustrate the results by appropriate diagrams.

1. $A \cup B$
2. $A \cap B$
3. $\bar{A} \cup B$
4. $\bar{A} \cap B$
5. $A \cup \bar{B}$
6. $A \cap \bar{B}$
7. $\bar{A} \cup \bar{B}$
8. $\bar{A} \cap \bar{B}$

9. $\overline{A \cup B}$ 10. $\overline{A \cap B}$ 11. $\overline{A \cup \bar{B}}$ 12. $\overline{A \cap \bar{B}}$
13. $\overline{\bar{A} \cup B}$ 14. $\overline{\bar{A} \cap B}$ 15. $\overline{\bar{A} \cup \bar{B}}$ 16. $\overline{\bar{A} \cap \bar{B}}$
17. $\emptyset \cup A$ 18. $\emptyset \cap A$

If U is the set of all positive integers, A is the set $\{1, 2, 3, 4\}$, and B is the set $\{3, 4, 5, 6\}$, find the following sets. Draw appropriate diagrams.

19. $A \cup B$ 20. $A \cap B$ 21. $A \cap \bar{B}$ 22. $\bar{A} \cap B$
23. $\bar{A} \cap \bar{B}$ 24. $\overline{A \cap \bar{B}}$

If A is any subset of U, find:

25. $A \cup \bar{A}$ 26. $A \cap \bar{A}$ 27. $A \cup \bar{A}$

In any triangle ABC, let L_1 be the set of all points on the bisector of angle A, L_2 be the set of all points on the bisector of angle B, and L_3 the set of all points on the bisector of angle C. Describe the following sets.

28. $L_1 \cap L_2$
29. $L_2 \cap L_2$
30. $(L_1 \cap L_2) \cap L_3$
31. $L_1 \cap (L_2 \cap L_3)$
32. If we denote by U the set of points interior to triangle ABC, and I the set of points in Problem 31, describe the relation of I and U.
33. If ABC is a triangle with an obtuse angle, U the set of points in the interior of the triangle, and I the set $h_1 \cap (h_2 \cap h_3)$ where h_1, h_2, and h_3 are the altitudes from each vertex of the triangle, describe the relation of I and U.
34. Suppose sets A and B are disjoint sets, and sets B and C are also disjoint sets. Are sets A and C disjoint sets? Illustrate your answer with a diagram.
35. Let C be the set of points on a circle and T be a tangent drawn from the end point of a radius. Describe $C \cap T$.
36. Let C be the set of points on a circle and T be a tangent drawn from a point outside the circle. Describe $C \cap T$.
37. Let A be the set of all right triangles and B be the set of all isosceles triangles with two 45° angles. Describe $A \cup B$.
38. In Problem 37, describe $A \cap B$.

2

What is a Number System?

9. THE DEVELOPMENT OF A SYSTEM OF NUMERATION

Mathematics began with the invention of counting numbers, and these numbers have continued to play an important role in its development. Today, however, mathematics is concerned not only with the use of numbers in *computation*, but it is equally concerned with the *structure* and *concepts* of number systems. We will take note of both of these aspects of mathematical thought in our development of number systems.

Numeration by One-to-One Correspondence

The first need for a system of numeration undoubtedly stemmed from a utilitarian need for counting. Obviously, counting offers no practical difficulty if it is limited to reasonably small quantities. Primitive man quickly found that tally marks made on the ground or wall of his cave, knots tied in a cord, pebbles accumulated in a pile, or notches cut into a stick were adequate to establish a sense of correspondence between physical objects. For example, he could establish a correspondence between the sheep in his flock and the pebbles in a pile. Today we consider each of these groupings as a set. If the set of sheep and the set of pebbles could be paired in such a way that to each member of the first set there was paired a single member of the second set, and to each member of the second set there was paired a single member of the first set, then we say that a one-to-one correspondence of the two sets was established. At the end of a season, if the set of sheep and the set of pebbles could not be put into one-to-one correspondence, then either some sheep were missing (evidenced by a surplus of pebbles) or there had been an increase in the flock (evidenced by a surplus of sheep). Thus, without resorting to formal counting, primitive man could check the size of his flock.

The device of placing objects into one-to-one correspondence with convenient "markers" is of more than passing importance. Many modern illustrations of the use of this concept can be cited. For example, in certain statistical procedures a simple distribution is made by tallying, that is, by setting up a one-to-one correspondence between tally marks and, let us say, raw scores that are under investigation. Similarly, when tickets are printed for reserved seats for a performance, a one-to-one correspondence is established between the printed tickets and the available seats. Modern mathematics, too, utilizes the concept of one-to-one correspondence (as was previously noted) to demonstrate that the number of points on any given line segment is equivalent to the number of points on any other line segment, or that the number of odd numbers is equivalent to the number of natural numbers.

Numbers and Numerals

Long after man began to identify quantity by means of correspondence, he began to separate his markers into groups of uniform size. This was a predictable response to a growing need for dealing with larger quantities. Some ancient tribes appear to have used groups of two; others used groups of three, or five, or even larger groupings to make up basic counting sets. A group of five sheep, or five fish, or five spears could each be placed in one-to-one correspondence with a basic set of markers, such as a set of five stones. It gradually became apparent that there was a property common to each of these groups which was quite independent of the objects themselves. The realization of this abstraction was the beginning of the number concept.

Today, the numbers associated with these basic sets of markers are known as *natural numbers* because they have been considered as having an existence quite independent of man. Man's role was to recognize that a property of number is attached to various groupings and to invent representations for it. The symbols 1, 2, 3, 4, 5, ... are our modern representation for the natural numbers, but many other representations have been used in the course of history. Among these are the Roman numerals I, II, III, IV, V, ... , the Greek symbols $\alpha, \beta, \gamma, \delta, \varepsilon, \ldots$, and the Egyptian notation |, ||, |||, ||||, |||||, ... Each of these symbols is properly called a *numeral* rather than a *number*. The essential distinction is that number is an abstract concept—a property shared by sets that can be placed into a one-to-one correspondence, while a numeral is a symbol used to represent a number. We shall make this distinction whenever it is helpful in understanding the discussion.

The Decimal System of Numeration

The familiar set of numerals 1, 2, 3, 4, 5, ... constitutes a *system of numeration* in that it introduces a basic set of symbols, together with rules for combining these symbols to name numbers. It is known as the Hindu-Arabic system because the notation was invented in India and introduced into the Western world by the Arabs. As everyone knows, if he is reminded of it, it is based on groupings of ten—therefore it is also known as a *decimal system* (from the Latin word *decem*, meaning ten).

Our decimal system is so much a part of our daily lives that it is difficult to appreciate how compact and sophisticated it is. We know, of course, that the symbol 374 means 3 one-hundreds, 7 tens, and 4 ones, and that the one-hundreds were a result of regrouping ten groups of ten. Further, we would readily agree that this representation is superior to, let us say, the Roman system designation

$$\text{CCCLXXIV}$$

which means $100 + 100 + 100 + 50 + 10 + 10 +$ (1 less than 5). Finally, no one would seriously question the computational advantages of the symbol 374 over the equivalent numeral CCCLXXIV.

But none of this really gets at the structure and concepts of our system of numeration; indeed, it is difficult to do so because we cannot easily take a detached view of something so familiar. In order to develop an objective examination of our decimal system, we will now consider one of many possible different but parallel systems of numeration.

The Hexal System of Numeration

Our decimal system is said to have *base* ten, a development that is completely arbitrary and probably just an anthropological accident. Man has ten fingers on his hands and, therefore, found it convenient to count in groups of ten. It is possible to count in groups of any natural number greater than one. For example, if for some strange reason early man had chosen to count in groups of six, we would today have a perfectly viable number system with base six. Such a system of numeration is called a *hexal system*.

A hexal system of numeration involves the use of just six symbols. We choose the familiar symbols 0, 1, 2, 3, 4, 5 with their common meanings, since nothing is gained by assigning new representations for these numerals.

What is a Number System?

Thus the word "one" and the symbol "1" are associated with any set that can be placed in one-to-one correspondence with the set $\{*\}$, the word "two" and the symbol "2" are associated with any set that can be placed in one-to-one correspondence with the set $\{*,\#\}$, and so on through the symbol 5. Now consider the following set of elements:

$$\{*,\#,\Delta,\alpha,X,\Sigma\}$$

How can we designate the number of elements in this collection? Clearly, we can no longer use the symbol "6" since we are limited only to the symbols 0,1,2,3,4,5. Since we have exhausted our one-digit numerals, we use two digits for the numeral associated with these elements. The smallest two-digit numeral we can form will be written (for the present) as

1,0

which is read "one comma zero" and means one group of six and no units.

Following this scheme, the numeral 1,1 would mean one group of six and one unit (or seven elements in our decimal system), and 1,2 would mean one group of six and two units (or eight elements in our decimal system). Similarly, 4,3 would mean four groups of six and three units, or 27 in our decimal system, and 4,5 would translate to 29 in our decimal system.

The development of a counting sequence in the hexal system can be summarized in the following listing.

Base 6 Numeral	Interpretation	Equivalent Base 10 Numeral
1,0	1 six and no units	6
1,1	1 six and 1 unit	7
1,2	1 six and 2 units	8
1,3	1 six and 3 units	9
1,4	1 six and 4 units	10
1,5	1 six and 5 units	11
2,0	2 sixes and no units	12
2,5	2 sixes and 5 units	17
3,0	3 sixes and no units	18
3,5	3 sixes and 5 units	23
4,0	4 sixes and no units	24
4,5	4 sixes and 5 units	29
5,0	5 sixes and no units	30

The largest two-digit numeral in the system is 5,5, which represents 5 groups of six and 5 units, or a total of 35 elements in our decimal system. It is evident that the next larger number in the hexal sequence will require three digits for its representation. The smallest such number is 1,0,0, which therefore represents one group of 36, no groups of six, and no units. We have now established a three-position hexal system where the place values are 36, six, and one, respectively. Thus, the hexal numeral 1,3,2 represents 1 group of 36, 3 groups of six, and 2 units, or a total of 56 elements in the decimal system. As before, the largest three-digit hexal numeral is 5,5,5 which is 215 in the decimal system. Therefore, the smallest four-digit hexal numeral is 1,0,0,0 which represents 216 in the decimal system. The pattern of development of place values in the hexal and decimal systems is apparent in the comparison shown in Figure 2.1.

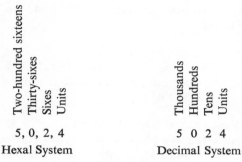

Figure 2.1

The relationship of place values in both the decimal and hexal systems is even more striking if these values are expressed in exponential notation. We will accept, *by agreement*, that 216, which is $6 \times 6 \times 6$, can be written as 6^3 and, similarly, that 36, which is 6×6, can be written as 6^2. We will further agree that the symbols 6^1 and 6 mean the same thing and that the symbol 6^0 represents the number 1. Thus, the base 6 numeral 5,0,2,4 can now be written in an *expanded notation* as follows:

$$5,0,2,4 = (5 \times 6^3) + (0 \times 6^2) + (2 \times 6^1) + (4 \times 6^0)$$

By an analogous development, the base 10 numeral 5024 can be written in expanded notation as

$$5024 = (5 \times 10^3) + (0 \times 10^2) + (2 \times 10^1) + (4 \times 10^0)$$

Up to this point we have been using commas to separate the digits of a hexal numeral, first, to guard against reading it as a decimal numeral and, second, to emphasize the place value of each digit according to its position. Now that we have introduced these ideas to caution the reader properly, we proceed to a customary and somewhat more sophisticated notation. The base 6 numeral 3,2,1,4 is generally written 3214_6 and is read "three, two, one, four, base 6." Note that the subscript is usually written in terms of base 10 numerals. Using this notation, the relationship of the hexal number 3214_6 to its decimal equivalent can be expressed as follows:

$$3214_6 = (3 \times 6^3 + 2 \times 6^2 + 1 \times 6^1 + 4 \times 6^0)_{10} = 730_{10}$$

EXAMPLE 1. Express 452_6 as a decimal number.
Solution: $452_6 = (4 \times 6^2 + 5 \times 6 + 2)_{10} = 176_{10}$.

EXAMPLE 2. Which is larger, 452_6 or 1001_6?
Solution: 1001_6 is greater than 216_{10}, since the first digit of 1001_6 represents a place value of 216_{10}. 452_6 is less than 216_{10}. (See Example 1.) Therefore, $1001_6 > 452_6$. Note that, as with decimal numbers, the one having more places is the larger.

EXAMPLE 3. Which is larger, 452_6 or 542_6?
Solution: $452_6 = 176_{10}$; $542_6 = 206_{10}$. Therefore, $542_6 > 452_6$. Note that, as with decimal numbers, if both have the same number of digits, the one having the larger digit farthest to the left is the larger.

EXAMPLE 4. Express 176_{10} as a hexal number.
Solution: We know that the place values of the hexal system of numeration are ... 6^3, 6^2, 6^1, 6^0. Since $176 < 6^3$, we have no digit in the 6^3 place, that is, we do not need four digits in our representation. Since $176 > 6^2$, we have a digit in the 6^2 place. Therefore our base 6 number will have three places. The first digit is 4 because 6^2 or 36 is contained in 176 four times, with a remainder of 32. The next digit is 5 because 6 is contained in 32 five times, with a remainder of 2. The last digit is 2, and the required number is 452_6.

We note that the hexal system of numeration has features strongly suggestive of the decimal system. We can "count," that is, we can set up a correspondence between the elements of any given set and one of a sequence of hexal numbers; we have place values associated with the position of each digit; we employ a zero symbol as a digit to denote "no units" in a given place; and we can tell at a glance whether one number is larger than another. But, although these are important virtues, a place-value system of representation has two additional advantages which are noteworthy.

1. Given the representation for any two natural numbers a and b, we can easily compute the representation of $a + b$.

2. Given the representation for any two natural numbers a and b, we can easily compute the representation of $a \times b$.

The key to these fundamental computations is the set of basic addition and multiplication facts. For the hexal system of numeration, the basic relations are summarized in Tables 2.1 and 2.2.

9. The Development of a System of Numeration

Table 2.1 Hexal Addition

+	1	2	3	4	5
1	2	3	4	5	10
2	3	4	5	10	11
3	4	5	10	11	12
4	5	10	11	12	13
5	10	11	12	13	14

Table 2.2 Hexal Multiplication

×	1	2	3	4	5
1	1	2	3	4	5
2	2	4	10	12	14
3	3	10	13	20	23
4	4	12	20	24	32
5	5	14	23	32	41

Calculations using the base 6 system may be carried out in a manner similar to that used with decimal numbers. The following examples illustrate the procedure.

EXAMPLE 5. Find the sum $(324 + 451 + 213)_6$.

Solution: *Step 1.* Adding the units column and referring to Table 2.1, we have $4 + 1 = 5$, $5 + 3 = 12$. Therefore, the sum of the units column is 1 six and 2 units. We write the 2 in the units column and consider the 1 to be a digit in the sixes column.

$$\begin{array}{c} \text{Base 6} \\ 324 \\ 451 \\ 213 \\ \hline 1432 \end{array}$$

Step 2. Adding the second column, we obtain 13 sixes. Therefore, we write the 3 in the sixes column and consider the 1 to be a digit in the thirty-sixes column.

Step 3. The sum of the digits in the thirty-sixes column is 14, which is written as 1 two-hundred-sixteen and 4 thirty-sixes.

$$\begin{array}{ccc} \text{Base 6} & & \text{Base 10} \\ 324 & \longrightarrow & 124 \\ 451 & \longrightarrow & 175 \\ 213 & \longrightarrow & 81 \\ \hline 1432 & \longrightarrow & 380 \end{array}$$

We can check our work by converting the problem to the decimal system. The sum of the decimal equivalents is 380. We note that $1432_6 = 380_{10}$. Therefore, the addition has been verified.

What is a Number System?

EXAMPLE 6. Using Table 2, find the product $(43 \times 25)_6$.

Solution: Step 1. Multiplying by 5, we have $5 \times 3 = 23$ units, and $5 \times 4 = 32$ sixes, which is equivalent to 34 sixes and 3 units, or 3 thirty-sixes, 4 sixes, and 3 units.

$$\begin{array}{r} \textit{Base 6} \\ 43 \\ 25 \\ \hline 343 \\ 130 \\ \hline 2043 \end{array}$$

Step 2. Multiplying by 2 sixes, we have 2 sixes $\times 3 = 10$ sixes, and 2 sixes $\times 4$ sixes $= 12$ thirty-sixes, which is a total of 1 two-hundred-sixteen, 3 thirty-sixes, and 0 sixes.

Step 3. Adding the partial products, we obtain the base 6 product 2043. Converting each of the base 6 factors to its equivalent decimal number, the problem is converted to the product $(27 \times 17)_{10}$, or 459_{10}. Since $2043_6 = 459_{10}$, the multiplication checks.

$$\begin{array}{r} \textit{Base 10} \\ 27 \\ 17 \\ \hline 189 \\ 27 \\ \hline 459 \end{array}$$

EXERCISE 9

THE DEVELOPMENT OF A SYSTEM OF NUMERATION

Which of the following pairs of sets can be placed in one-to-one correspondence?

1. Families and cars.
2. Trees and tree trunks.
3. Sodium atoms and chlorine atoms in table salt.
4. Planets and planetary orbits.
5. Even numbers and odd numbers.
6. Lines of latitude and lines of longitude.
7. Circles and diameters of circles.
8. Spheres and centers of spheres.
9. Natural numbers and "perfect squares," that is, numbers such as 1, 4, 9, ...
10. Nouns and verbs as they appear in sentences.

Given the Roman numerals I = 1, V = 5, X = 10, L = 50, C = 100, M = 1000, write the Roman numeral for each of the following expressions.

11. 257
12. 94
13. 1966
14. 1492
15. XVI + IV
16. LXVII + CLXXVI
17. XCLVII + XVIII
18. XLI + XCVIII
19. The product of XXVIII and VII
20. The product of LXVII and XVI

Using base 6 numeration:

21. Set up a calendar for the month of January of this year.
22. Set up a calendar for the month of February of this year.

Write each of the following hexal numerals in expanded decimal notation and evaluate as illustrated in Problem 23.

23. $2354_6 = (2 \times 6^3 + 3 \times 6^2 + 5 \times 6^1 + 4 \times 6^0)_{10} = 574_{10}$
24. 1042
25. 432
26. 1001
27. 354
28. 234
29. 412
30. 241
31. 555
32. 304
33. (a) What is the decimal notation for the numeral 5555_6?
    (b) What, therefore, is the decimal notation for the numeral 10000_6?
    (c) What is the decimal notation for the numeral $111,111_6$?
34. (a) Write the numeral for one dozen in the hexal system.
    (b) Write the numeral for one gross in the hexal system.
    (c) Verify that in the hexal system one gross is equal to one dozen dozen.

Express each of the following decimal numerals as a hexal numeral.

35. 75
36. 135
37. 273
38. 64
39. 185
40. 391

Perform each of the following operations in the hexal system and check by translating the problem into the decimal system.

41. 14 + 25
42. 23 + 34
43. 115 + 542
44. 503 + 324
45. 42 + 35 + 22
46. 53 + 14 + 21
47. 1123 + 4315
48. 3502 + 4041
49. 34 × 25
50. 43 × 15
51. 134 × 21
52. 453 × 32
53. 103 × 4
54. 531 × 42
55. 305 × 204

10. NUMERATION FOR COMPUTERS

Although any number greater than one can be used as the base of a system of numeration, certain nondecimal systems are of particular importance because of their use in modern computers. To understand such use of numeration, a few comments about computers will be helpful.

The automatic computer follows such devices as the abacus, slide rule, and calculating machine in the evolution of man's efforts to solve his mathematical problems. It differs from these devices in two important respects—it solves significant problems at fantastic speeds, and it solves them without the need of human intervention during the course of solution. These are profound differences, and they enable computers to take over control of factories, solve complex research problems, make instantaneous decisions based on rapidly changing information, help design and test simulated operations, and process a multitude of data forms.

There are two basic types of computers, the digital computer, which is essentially a counting device, and the analog computer, which is essentially a measuring device. Our discussion will be limited to digital computers. It is important to recognize, at the onset, that a digital computer is not merely a high-speed electronic version of a desk calculator. Desk calculators deal with decimal numbers directly through the use of wheels which have 10 discrete positions. Although many 10-state devices are possible, few have been found that are fast enough and reliable enough for digital computers. The most practical devices for representing numbers and instructions in a computer have proved to be 2-state or *binary* devices. Because these involve only two symbols—"0" and "1"—they are able to depict such conditions as on or off, conducting or nonconducting, yes or no, open or closed, and so on. There is no intermediate condition and, therefore, the unreliability resulting from errors and indecision is minimal. The implication of this is that a number system for computer use is ideally limited to just two symbols. We shall now consider such a system.

Binary Notation

A system of numeration that employs only two symbols is called a *binary system*. It is based on powers of 2, just as the decimal system, which employs ten symbols (0,1,2,3,4,5,6,7,8,9), is based on powers of 10, and the hexal system, with six symbols, is based on powers of 6. In practice, the two symbols used in the binary system are 0 and 1. That every natural number is expressible as a sum of powers of 2 is suggested by the following illustration.

Suppose we wish to represent the decimal number 93 as a sum of powers of 2. We note that, as a consequence of previous agreements, $2^0 = 1$, $2^1 = 2$, $2^2 = 4$, $2^3 = 8$, $2^4 = 16$, $2^5 = 32$, $2^6 = 64$, $2^7 = 128$, and so on. Since the decimal number 93 falls between 2^6 and 2^7, then

$$93 = 2^6 + 29$$

Now the remainder, 29, falls between 2^4 and 2^5, so that
$$93 = 2^6 + 2^4 + 13$$
Continuing in this manner, we finally obtain
$$93 = 2^6 + 2^4 + 2^3 + 2^2 + 2^0$$
Therefore, in expanded notation we have
$$93 = (1 \times 2^6) + (0 \times 2^5) + (1 \times 2^4) + (1 \times 2^3)$$
$$+ (1 \times 2^2) + (0 \times 2^1) + (1 \times 2^0)$$

Following the procedure used in both the decimal and hexal systems, we need record only the digit in each position of the numeral, taking care to specify the base of the system. Then
$$93_{ten} = 1{,}011{,}101_2$$

We can summarize the position value of each digit in the binary system by means of Table 2.3 for natural numbers.

Table 2.3 Binary System Place Values

Position (counting from right)	nth	⋯	7th	6th	5th	4th	3rd	2nd	1st
Position value as power of 2	2^{n-1}	⋯	2^6	2^5	2^4	2^3	2^2	2^1	2^0
Position value as decimal numeral		⋯	64	32	16	8	4	2	1

EXAMPLE 1. Represent $10{,}011{,}101_2$ as a decimal numeral.
Solution: In expanded notation we have
$$10{,}011{,}101_2 = (1 \times 2^7 + 1 \times 2^4 + 1 \times 2^3 + 1 \times 2^2 + 1 \times 2^0)_{ten}$$
$$= (128 + 16 + 8 + 4 + 1)_{ten} = 157_{ten}$$

EXAMPLE 2. Express 51_{ten} as a binary number.
Solution: The decimal number 51 falls between 2^6 which is 64 and 2^5 which is 32. Therefore
$$51 = 2^5 + 19$$
$$= 2^5 + 2^4 + 3$$
$$= 2^5 + 2^4 + 2^1 + 2^0$$

Referring to the place value system for binary numbers, we have
$$51_{ten} = 110{,}011_2$$

The operation of addition or multiplication of two binary numbers is astonishingly simple, since it is based on Tables 2.4 and 2.5.

64   What is a Number System?

Table 2.4 Binary Addition   **Table 2.5** Binary Multiplication

+	0	1
0	0	1
1	1	10

×	0	1
0	0	0
1	0	1

EXAMPLE 3. Perform the operation 100,010 + 10,011 in the binary system and then perform the equivalent operation in the decimal system.
Solution:

Base 2		*Base 10*
100010	$\longleftrightarrow$	34
10011	$\longleftrightarrow$	19
110101	$\longleftrightarrow$	53

EXAMPLE 4. Find the product of the binary numbers 1101 and 111, then check by converting the problem to decimal numeration.
Solution:

Base 2   *Base 10*

1101 $\longleftrightarrow$ 13
 111 $\longleftrightarrow$ 7

1101   91
1101
1101

1011011 $\longleftrightarrow$ $(64 + 16 + 8 + 2 + 1)_{ten} = 91_{ten}$

Binary Coded Systems of Numeration

Since computers operate in a binary code but our mathematical transactions are generally performed in a decimal system, it is not unexpected that the binary system is adapted to represent decimal numbers directly. The problem admits of a simple solution: four places on the computer are used to represent any one decimal digit, and the conversion is effected as follows:

Base 10 digit	0	1	2	3	4	5	6	7	8	9
Four-place binary notation	0000	0001	0010	0011	0100	0101	0110	0111	1000	1001

10. Numeration for Computers

Thus the decimal number 3907 is represented as

$$3907_{ten} = (0011\ \ 1001\ \ 0000\ \ 0111)_{bcd}$$

where "bcd" means "binary coded decimal" notation. Conversely, it is simple to interpret a bcd numeral as a decimal numeral, for example,

$$(0101\ \ 0100\ \ 0110)_{bcd} = 546_{ten}$$

Although the bcd system is an interesting example of a combination of decimal and nondecimal notation, it is not an efficient system for computer operations because it utilizes only 10 of the 16 possible values that can be assigned to a four-place group of binary digits. To overcome this deficiency, some computers are designed for a *hexadecimal system* of notation—a number system using the base 16. Another common variation is to design the computer to use an *octal system* of notation (a number system using the base 8), representing each octal digit by a three-place group of binary digits. Some mathematical aspects of octal notation will be considered in the problems that follow.

EXERCISE 10

NUMERATION FOR COMPUTERS

Express each of the following as a decimal numeral.

1. 101_2
2. 10101_2
3. 11011_2
4. $(1001\ 0101\ 0111\ 0010)_{bcd}$
5. $(0110\ 0101\ 0001)_{bcd}$
6. $(1001\ 0101\ 0110)_{bcd}$
7. $(0111\ 0101\ 1000\ 1001)_{bcd}$

Express each of the following decimal numerals as a base 2 numeral.

8. 17
9. 36
10. 13
11. 29
12. 85
13. 91
14. 92
15. 127

Express each of the following decimal numerals in bcd notation.

16. 73
17. 96
18. 149
19. 228
20. 1037
21. 7142
22. 8099
23. 2763

Perform the indicated operations in the base 2 system and check by converting into the decimal system.

24. $101_2 + 11_2$
25. $1110_2 + 1010_2$
26. $11011_2 + 1011_2$
27. $1101_2 \times 11_2$
28. $1011_2 \times 10_2$
29. $1011_2 \times 101_2$

Expand each of the given octal numerals and express in decimal notation, as illustrated in Problem 30.

30. $273_8 = (2 \times 8^2 + 7 \times 8^1 + 3 \times 8^0)_{ten} = (128 + 56 + 3)_{ten} = 187_{ten}$

66   *What is a Number System?*

31. 63_8 **32.** 75_8 **33.** 206_8 **34.** 777_8
35. 570_8 **36.** 307_8 **37.** 621_8 **38.** 1111_8

Express each of the given octal numerals in a three-place binary coded octal notation (bco), as illustrated in Problem 39.

39. $513_8 = (101\ 001\ 011)_{bco}$ **40.** 275_8
41. 1432_8 **42.** 2760_8 **43.** 763_8

44. The following diagrams represent displays of numbers in the register of a computer using the binary coded decimal system. Read the number if the value assigned to each shaded circle reading upward is 1, 2, 4, 8, respectively.

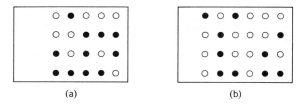

(a)   (b)

11. MATHEMATICAL SYSTEMS

The natural numbers, together with certain operations and axioms, constitute a good example of a *mathematical system*. In general, a mathematical system is described by specifying a set of elements, defining one or more operations or relations for members of the set, and stating the axioms that the elements, operations, and relations satisfy. If the set of elements under consideration is a finite set, the system is known as a *finite mathematical system*. As an introduction to the treatment of natural numbers as a mathematical system, we will examine several examples of finite systems.

A Finite System of Natural Numbers

A subset of the natural numbers generally provides a simple example of a finite system. Let us consider the subset $A = \{1,2,3,4,5,6,7,8,9,10,11,12\}$, together with the operations of addition and multiplication. Let us also accept, for the time being, an intuitive interpretation of these operations—that each is a binary operation on the set A which produces a unique result when acting on any two given elements of the set. We denote the operation of addition by the symbol "+." The given elements of the set are called *addends* or *terms* and the result of the operation is called the *sum*. The basis of the operation is counting—for example, the sum of 3 and 4 is determined by combining a set of three elements with a set of four elements (without common members) and counting the members of the resulting set.

We note that in this simplified system the operation of addition is not always possible. For example, there is no element in the set A which is equal to $6 + 8$. We say that the set A is *not closed* under the operation of addition, that is, the sum is not always an element in A.

Similarly, the operation of multiplication, denoted by the symbol "$\times$," can be interpreted as a process of combining a collection of like addends. Thus, 4×3 is determined by counting all the elements of three sets, each having four elements without common members. The result of the operation is called the *product*, and the numbers which are multiplied to form the product are called *factors*. It is evident that the set A is not closed under the operation of multiplication—for example, the product 5×9 is not contained in the set.

The system has an element, 1, which has a special property. If we multiply any element in set A by 1, that element remains unchanged. The element 1 is called an *identity element* for the operation of multiplication in the set A. Note that there is no identity element in set A for the operation of addition, that is, there is no element which, when added to a second element, always results in the second element.

A System of Clock Numbers

To consider some further properties of a simple mathematical system, we now arrange the 12 numbers in set A in the format of a clock (Figure 2.2) and call this collection set B. The set B, together with the operations of addition and multiplication, forms a new finite mathematical system with some interesting properties.

In this system, we define addition in a predictable way, $8 + 5$ being interpreted as a movement of the clock hand from an initial position at 12 to the numeral 8, then an additional movement of five places. Thus $8 + 5 = 1$.

Figure 2.2

Multiplication, too, may be defined in a natural way. For example, the product 7×3 may be interpreted as a seven-place movement of the clock hand performed three times. If we again assume the initial position of the clock hand at 12, then $7 \times 3 = 9$.

A complete definition of addition and multiplication in this system is given by listing all possible addition and multiplication relations, as shown in Tables 2.6 and 2.7.

Table 2.6 Addition

+	1	2	3	4	5	6	7	8	9	10	11	12
1	2	3	4	5	6	7	8	9	10	11	12	1
2	3	4	5	6	7	8	9	10	11	12	1	2
3	4	5	6	7	8	9	10	11	12	1	2	3
4	5	6	7	8	9	10	11	12	1	2	3	4
5	6	7	8	9	10	11	12	1	2	3	4	5
6	7	8	9	10	11	12	1	2	3	4	5	6
7	8	9	10	11	12	1	2	3	4	5	6	7
8	9	10	11	12	1	2	3	4	5	6	7	8
9	10	11	12	1	2	3	4	5	6	7	8	9
10	11	12	1	2	3	4	5	6	7	8	9	10
11	12	1	2	3	4	5	6	7	8	9	10	11
12	1	2	3	4	5	6	7	8	9	10	11	12

Table 2.7 Multiplication

×	1	2	3	4	5	6	7	8	9	10	11	12
1	1	2	3	4	5	6	7	8	9	10	11	12
2	2	4	6	8	10	12	2	4	6	8	10	12
3	3	6	9	12	3	6	9	12	3	6	9	12
4	4	8	12	4	8	12	4	8	12	4	8	12
5	5	10	3	8	1	6	11	4	9	2	7	12
6	6	12	6	12	6	12	6	12	6	12	6	12
7	7	2	9	4	11	6	1	8	3	10	5	12
8	8	4	12	8	4	12	8	4	12	8	4	12
9	9	6	3	12	9	6	3	12	9	6	3	12
10	10	8	6	4	2	12	10	8	6	4	2	12
11	11	10	9	8	7	6	5	4	3	2	1	12
12	12	12	12	12	12	12	12	12	12	12	12	12

The following properties of the system are immediately apparent.

1. The set B is closed with respect to the operation of addition, that is, the sum of any two element of set B is always an element of set B.

2. The set B is closed with respect to the operation of multiplication.

3. The element 12 is an identity element for addition, that is, if 12 is added to any element, the sum is that element.

4. The element 1 is an identity element for multiplication.

We note further that the tables display a symmetry of entries along the diagonal from upper left to lower right. This symmetry is a visual display of

11. Mathematical Systems

what is known as the *commutative property* of addition and multiplication. For example, $9 + 5 = 2$ and $5 + 9 = 2$. Therefore, $9 + 5 = 5 + 9$ and, in general, $a + b = b + a$ for any two elements in the system. Similarly, for multiplication, $5 \times 4 = 8$ and $4 \times 5 = 8$ or $5 \times 4 = 4 \times 5$. This is an example of the general property of commutativity for multiplication, that $a \times b = b \times a$ for any two elements in the system.

Another important property of the system, which is not easily apparent from an inspection of the tables, can be verified by checking some individual cases. It is the property of *associativity*. The general form of the associative property for addition is stated as

$$(a + b) + c = a + (b + c)$$

where a, b, and c are any three elements in the set B. Similarly, the associative property for multiplication is given by the relation

$$(a \times b) \times c = a \times (b \times c)$$

EXAMPLE 1. Show that the associative property holds for the addition of the elements 9, 5, 3, respectively.

Solution: We are required to show that

$$(9 + 5) + 3 = 9 + (5 + 3).$$

Referring to Table 2.6, we have
$$2 + 3 = 9 + 8$$
$$5 = 5$$

EXAMPLE 2. Verify the associative property $(4 \times 5) \times 2 = 4 \times (5 \times 2)$.
Solution:
$$(4 \times 5) \times 2 = 4 \times (5 \times 2)$$
$$8 \times 2 = 4 \times 10$$
$$4 = 4$$

The properties of commutativity and associativity relate to the operations of addition and multiplication taken separately. An additional important property relates the two operations. To illustrate this property, consider the expression

$$4 \times (5 + 6)$$

If we perform the operations in the order indicated by the parenthesis, the result is 4×11 or 8. The same result is obtained if we multiply 4×5, then 4×6, and add the products, since

$$4 \times 5 = 8$$
$$4 \times 6 = 12$$
$$8 + 12 = 8$$

In brief then, we have shown that

$$4 \times (5 + 6) = (4 \times 5) + (4 \times 6)$$

It can be said that the multiplication appears to distribute itself over the addition; mathematically we say that *multiplication is distributive with respect to addition*. In general, this property states that if a, b, and c are any three elements in set B, then

$$a \times (b + c) = (a \times b) + (a \times c)$$

EXAMPLE 3. Show that in set B addition is not distributive with respect to multiplication.

Solution: If addition were distributive with respect to multiplication, then for any three elements a, b, c in set B,

$$a + (b \times c) = (a + b) \times (a + c)$$

If we select, at random, $a = 2$, $b = 3$, $c = 5$, then

$$2 + (3 \times 5) \neq (2 + 3) \times (2 + 5)$$

because the left member is 5 and the right member is 11.

Example 4. Using the distributive property, rewrite the expression $3 \times (2 + 8)$.

Solution:

$$3 \times (2 + 8) = (3 \times 2) + (3 \times 8)$$
$$= 6 + 12$$
$$= 6$$

The system of clock numbers is also a useful vehicle for introducing certain properties of the familiar operations of subtraction and division. Subtraction is said to be the *inverse* of addition, and division is said to be the *inverse* of multiplication. The inverse of entering a room is leaving it; the inverse of pouring water into a container is to release water from the container. If we begin with the number a and add to it the number b, then subtract b from the sum, we return to the initial number a. Thus the operation of subtraction (denoted by the symbol "$-$") undoes the operation of addition and, similarly, the operation of division (denoted by the symbol "$\div$") undoes the operation of multiplication. Any two operations so related that one countereffects the other are said to be inverse operations.

If the general addition relation is written as $a + b = c$, then the corresponding subtraction relation is $c - b = a$, where c is called the *minuend*, b is called the *subtrahend*, and a is the *difference*. For example, the inverse of the particular addition relation $7 + 9 = 4$ is the subtraction relation $4 - 9 = 7$. That this relation is consistent with our usual physical interpretation of subtraction is evident because 9 hours preceding 4 o'clock is 7 o'clock.

Following this same line of reasoning, for each multiplication relation $a \times b = c$ in the system, there exists a corresponding division relation $c \div b = a$, where c is called the *dividend*, b is the *divisor*, and a is the quotient. Other methods of indicating division in common use are $c:b$, c/b, and $\frac{c}{b}$. Thus, since $4 \times 5 = 8$, then $8/5 = 4$.

EXAMPLE 5. What two subtraction relations result from the addition relation $8 + 5 = 1$?

Solution: The inverse relation of $8 + 5 = 1$ is $1 - 5 = 8$. Since addition is commutative, we can also write $5 + 8 = 1$, in which case the inverse relation is $1 - 8 = 5$. Therefore the two subtraction relations associated with the given addition relation are

$$1 - 5 = 8$$
$$1 - 8 = 5$$

EXAMPLE 6. Show all possible division relations (in the system of clock numbers) for which the dividend is 10.

Solution: Referring to Table 2.7, which shows the multiplication relations, we find in the first row the product 10 which is the result of the multiplication $1 \times 10 = 10$. Therefore, the corresponding division relations are $10 \div 10 = 1$ and $10 \div 1 = 10$. In the second row, we find $2 \times 5 = 10$ from which we obtain $10 \div 5 = 2$ and $10 \div 2 = 5$. There is no product 10 in the third or fourth rows (indicating that no division relation exists for $10 \div 3$ or $10 \div 4$). Continuing in this manner, we determine the complete list of the required division relations to be

$$10 \div 1 = 10$$
$$10 \div 2 = 5$$
$$10 \div 5 = 2$$
$$10 \div 7 = 10$$
$$10 \div 10 = 1$$
$$10 \div 11 = 2$$

In examining properties of subtraction and division that are analogous to those previously developed for addition and multiplication, we note the following.

1. The set of clock numbers is closed with respect to subtraction; it is not closed with respect to division.
2. Neither the operation of subtraction nor division is commutative.
3. Neither the operation of subtraction nor division is associative.

EXERCISE 11

MATHEMATICAL SYSTEMS

In the system of clock numbers (based on a 12-hour clock), show that each of the following equalities is true by evaluating both sides.

1. $8 + 5 = 5 + 8$
2. $9 + 11 = 11 + 9$
3. $(2 + 7) + 10 = 2 + (7 + 10)$
4. $(5 + 2) + 8 = 5 + (2 + 8)$
5. $7 \times (3 + 1) = (7 \times 3) + (7 \times 1)$
6. $4 \times (3 + 10) = (4 \times 3) + (4 \times 10)$
7. $4 \times 5 = 5 \times 4$
8. $(3 \times 7) \times 2 = 3 \times (7 \times 2)$
9. $(2 \times 4) \times 5 = 2 \times (4 \times 5)$
10. $9 \times 6 = 6 \times 9$

Which of the following are commutative operations?

11. Writing an examination and studying for an examination.
12. Mixing the paint and preparing the brush.
13. Setting an alarm clock and going to sleep.
14. Cleaning the windshield of a car and filling its tank.

Which of the following operations are associative?

15. Pipe plus tobacco plus lighted match.
16. Coffee plus cream plus stirring.
17. Coffee plus sugar plus cream.
18. Soil plus grass seed plus water.

Figure 2.3

Consider a 5-hour clock, as illustrated in Figure 2.3. Test the commutative law of addition and multiplication by verifying each of the following.

19. $3 + 4 = 4 + 3$
20. $5 + 2 = 2 + 5$
21. $4 + 2 = 2 + 4$
22. $2 \times 3 = 3 \times 2$
23. $5 \times 3 = 3 \times 5$
24. $4 \times 3 = 3 \times 4$

Use the 5-hour clock to test the associative or distributive laws of addition and multiplication by verifying each of the following.

25. $(1 + 3) + 4 = 1 + (3 + 4)$
26. $(4 + 2) + 5 = 4 + (2 + 5)$
27. $(3 + 2) + 2 = 3 + (2 + 2)$
28. $2 \times (3 \times 2) = (2 \times 3) \times 2$
29. $4 \times (1 + 5) = (4 \times 1) + (4 \times 5)$
30. $3 \times (2 + 4) = (3 \times 2) + (3 \times 4)$

Make a table showing all combinations for each of the following operations based on a 5-hour clock.

31. Addition. **32.** Subtraction.
33. Multiplication. **34.** Division.

Cite two examples to illustrate each of the following properties for a simple mathematical system based on a 5-hour clock.

35. The set is closed with respect to addition.
36. The set is closed with respect to multiplication.
37. Multiplication is distributive with respect to addition.
38. There exists an identity element for addition.
39. There exists an identity element for multiplication.

Given that a, b, and c are elements in the system of the 5-hour clock, and $\Box$ and $\bigcirc$ are symbols for any two of the operations of addition, subtraction, multiplication, and division. For what replacements of $\Box$ and $\bigcirc$ will the following statements be true?

40. $a \Box b = b \Box a$
41. $(a \Box b) \bigcirc c = a \Box (b \bigcirc c)$
42. $(a \bigcirc b) \bigcirc c = a \bigcirc (b \bigcirc c)$
43. $a \Box (b \bigcirc c) = (a \Box b) \bigcirc (a \Box c)$
44. $a \Box b \Box c = (a \Box b) \Box (a \Box c)$

45. A simple mathematical system which uses geometric transformations rather than numbers as its elements can be illustrated by constructing a rectangular card with its edges colored as indicated in Figure 2.4. We define its elements to be the following four rotations:

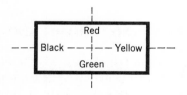

Figure 2.4

h = rotation of 1/2 revolution about the horizontal axis. (If we start with the position in Figure 2.4, the red and green edges will be interchanged.)
v = rotation of 1/2 revolution about the vertical axis. (Starting with the position in Figure 2.4, the black and yellow edges will be interchanged.)
r = rotation of 1/2 revolution about the center and in the plane of the rectangle. (Starting with the position in Figure 2.4, both pairs of opposite edges will be interchanged.)
i = rotation of 1 revolution about the center and in the plane of the rectangle. (No change results in the relative positions of the edges.)

The symbol $*$ is used to represent the operation "followed by." Thus, $r * h$ means that, starting with the initial position in Figure 2.4, we rotate through 1/2 revolution about the center to obtain the position illustrated in Figure 2.5, and follow this with a rotation of 1/2 revolution about the horizontal axis. The result is illustrated in Figure 2.6 which is the position we would obtain by the single rotation

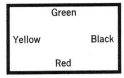

Figure 2.5

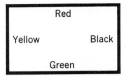

Figure 2.6

v. Therefore,
$$r * h = v$$

(a) Using this procedure, complete the following table for the operation $*$:

$*$	i	h	v	r
i				
h				
v				
r				

(b) Is the set of rotations closed with respect to the operation $*$? Be able to support your answer.
(c) Cite two relations to illustrate that the operation $*$ is or is not commutative.
(d) Cite two relations to illustrate that the operation $*$ is or is not associative.
(e) Does the set of rotations contain an identity element for the operation $*$? Be able to support your answer.

12. PEANO AXIOMS AND THE NATURAL NUMBERS

In the last section we considered certain subsets of natural numbers as finite mathematical systems, with particular reference to the meaning of the basic operations and their properties. We are now ready to develop the natural numbers as an infinite mathematical system, beginning with undefined terms and a set of axioms, defining the operations on the elements of the system, and using these axioms and definitions to derive other laws. The result will be a deductive system by means of which all the theorems of the arithmetic of natural numbers can be developed. It was formulated in the late nineteenth century by Giuseppe Peano, an Italian mathematician and logician.

The undefined terms of the system are "1," "number," and "successor." From a strictly logical viewpoint, these undefined terms admit of a variety of interpretations that will result in true propositions; however, it will guide our thinking to interpret them in a natural way. Therefore, we will agree that "1" is intended to name the number one, and "number" is intended to refer to one of the natural numbers. We will agree also that the natural

12. Peano Axioms and the Natural Numbers

numbers form a sequence and that the term "successor" refers to the number following a given number in the sequence. These interpretations will satisfy the five axioms that Peano listed to form the framework of his system. The five axioms may be stated as follows.

1. 1 is a number.
2. With each number n, there is associated uniquely another number n' which is called the successor of n.
3. 1 is not the successor of any number.
4. No two numbers have the same successor.
5. If S is a set of numbers such that (a) 1 is in S, and (b) the successor of each member of S is in S, then all the numbers are in S.

The first step in the construction of a mathematical system based on these axioms is the definition of the various natural numbers. Axiom 1 in combination with Axiom 3 defines the special position of 1 as the first natural number. The number 2 is then defined as the successor of 1, that is, $2 = 1'$. Similarly, 3 is defined as $2'$, 4 as $3'$, 5 as $4'$, and so on. Axiom 2 tells us that this process can be continued indefinitely, while Axiom 4 tells us that the process will never lead back to one of the numbers defined previously. Thus, Axioms 1 to 4 (together with our interpretations of the undefined terms) are sufficient to assure us that *only* natural numbers will be generated by the process of successors.

This, of course, is not the same as saying that *all* natural numbers will be so generated. Axiom 5 is needed to state explicitly that this last requirement is met. For example, if we wish to verify that any particular number, say, 117, is contained in the set of natural numbers, we merely start with the fact that 1 belongs to S by (a), and 2 (which is equal to $1'$) belongs to S by (b). Similarly, 3 (which is equal to $2'$) belongs to S by (b), and we have only to repeat the procedure a sufficient number of times to reach the conclusion that 117 belongs to S. Obviously, the same form of argument can be used with respect to any number n and, therefore, the effect of Axiom 5 is to assure that *each and every* natural number will be generated by the process of successors.

The fifth of Peano's axioms is called the *principle of finite induction*. Although the word "induction" is used in its name, the process should not be confused with inductive inference as discussed in Section 1. Inductive inference, we saw, was basically a method of discovering general results from individual experiences. The principle of finite induction, on the other hand, is a rigorous method of proof which is extremely useful in establishing certain kinds of mathematical statements. It can be likened to the problem of entering an unending sequence of connected rooms, assuming that each room contains a key that will unlock the door to the next room. Now, if we

76   *What is a Number System?*

can obtain a key to any one room, we can pass successively into all the remaining rooms. Specifically, if we have a key to the first room, we can use it to obtain the key to the second room, use the key found in that room to unlock the third room, and so on. Theoretically, we could enter *every* room in the entire complex.

Returning now to Peano's fifth axiom, we note that statement (a) corresponds to our having a key to the first room. We use this statement to establish that the successor of the number 1 is in set S, which is analogous to our finding the key to the second room. Just as each room yields a key that leads to the next room, so the repeated use of (b) generates a natural number which leads to the next natural number. Our conclusion that we could enter every room corresponds to the conclusion that we can generate *every* natural number by the process of successors.

Defining Operations on the Set of Natural Numbers

We will designate by the symbol N the set of natural numbers described by Peano, that is,

$$N = \{x \mid x \text{ is a natural number}\}$$

These are the elements of our mathematical system. Our next step is to define the operations on these elements. Addition is defined in the following manner, which expresses in a rigorous form the fact that addition is basically effected by the repeated addition of 1.

Definition 1. Addition of the natural numbers is a binary operation designated by the symbol "$+$" such that, for any two natural numbers n and k,

$$(1) \; n + 1 = n'$$
$$(2) \; n + k' = (n + k)'$$

EXAMPLE 1. Using definition 1, derive the relation $3 + 2 = 5$.
Solution:

$$\begin{aligned} 3 + 2 &= 3 + 1' & &\text{Definition of the numeral 2, } (2 = 1') \\ &= (3 + 1)' & &\text{Definition of addition} \\ &= (3')' & &\text{Definition of addition} \\ &= 4' & &\text{Definition of the numeral 4} \\ &= 5 & &\text{Definition of the numeral 5} \end{aligned}$$

The multiplication of two numbers may now be expressed in terms of addition in such a way that the idea of a product of two numbers is treated essentially as a sum of equal terms.

Definition 2. Multiplication of the natural numbers is a binary operation designated by the symbol "$\times$" or "$\cdot$" such that for any two natural numbers n and k,

(1) $n \cdot 1 = n$
(2) $n \cdot k' = n \cdot k + n$

EXAMPLE 2. Show that $7 \cdot 3 = 21$.
Solution: We assume that the numerals 1, 2, 3, ... 21 have been defined by the process of successors and that any particular addition relation can be derived by the procedure in Example 1. Then:

$7 \cdot 3 = 7 \cdot 2'$	Definition of the numeral 3
$= 7 \cdot 2 + 7$	Definition of multiplication
$= 7 \cdot 1' + 7$	Definition of the numeral 2
$= (7 \cdot 1 + 7) + 7$	Definition of multiplication
$= (7 + 7) + 7$	Definition of multiplication
$= 14 + 7$	Addition
$= 21$	Addition

The concept of subtraction and division as inverses of addition and multiplication, respectively, is defined as follows.

Definition 3. Subtraction on set N is an operation for which $c - b = a$ if and only if $a + b = c$ where a, b, and c are natural numbers.

Definition 4. Division on the set N is an operation for which $c \div b = a$ if and only if $a \cdot b = c$, where a, b, and c are natural numbers.

EXAMPLE 3. If $9 - 5 = a$, then $a + 5 = 9$, and if $a + 5 = 9$, then $9 - 5 = a$. In brief, we can express the same idea by means of the biconditional $9 - 5 = a \leftrightarrow a + 5 = 9$ which, of course, is true for the replacement $a = 4$.

EXAMPLE 4. $7 - 9 = a$ if and only if $a + 9 = 7$. Since there is no element in the set N for which the relation $a + 9 = 7$ is true, the set N is not closed with respect to subtraction.

EXAMPLE 5. $13 \div 4 = a$ if and only if $4 \cdot a = 13$. Since this statement is false for each $a \in N$, the set N is not closed with respect to division.

Properties of Addition and Multiplication

The familiar laws governing addition and multiplication of natural numbers can be derived solely on the basis of the Peano axioms and the definitions we

Table 2.8

Laws for Natural Numbers	Addition	Multiplication
Closure laws. For every pair of natural numbers, the result of the given operation is a unique natural number.	$a + b = c$	$a \cdot b = c$
Commutative laws. The result of the given operation is not affected by the sequential order of the elements.	$a + b = b + a$	$a \cdot b = b \cdot a$
Associative laws. The result of the given operation on three elements is not affected by different groupings.	$(a + b) + c = a + (b + c)$	$(a \cdot b) \cdot c = a \cdot (b \cdot c)$
Distribution law. Multiplication is distributive with respect to addition.	$a \cdot (b + c) = a \cdot b + a \cdot c$	

have listed. It is important to recognize that this is so, even though the actual proofs are beyond the scope and purpose of this book. Table 2.8, in which N is the set of natural numbers and a, b, and c are any elements in N, is a summary of these laws.

Our interest in these properties of the natural numbers is based on their importance in the understanding of the fundamental arithmetic processes as well as their role in the development of the theoretical structure of arithmetic. Although they may seem obvious, a little reflection will convince us that they are not obvious to the novice in arithmetic. For example, the preschool child who learns that 2 more than 9 is 11 is not aware, initially, that 9 more than 2 is also 11. It is only after considerable experience that the generalization which we know as the commutative law becomes apparent.

Similarly, the addition of three numbers, such as $4 + 7 + 5$, offers no practical difficulty—yet few are aware that the result can be obtained by any one of 12 different groupings of the numbers. That these 12 results are equivalent is a consequence of the associative and commutative laws. To illustrate, consider the sum $a + b + c$, where a, b, and c are any three natural numbers. The following development is a proof of the equivalence of these various groupings:

$$
\begin{aligned}
(a + b) + c &= a + (b + c) && \text{By the associative law} \\
&= a + (c + b) && \text{By the commutative law} \\
&= (a + c) + b && \text{By the associative law} \\
&= (c + a) + b && \text{By the commutative law} \\
&= c + (a + b) && \text{By the associative law} \\
&= c + (b + a) && \text{By the commutative law} \\
&= (c + b) + a && \text{By the associative law} \\
&= (b + c) + a && \text{By the commutative law} \\
&= b + (c + a) && \text{By the associative law} \\
&= b + (a + c) && \text{By the commutative law} \\
&= (b + a) + c && \text{By the associative law}
\end{aligned}
$$

The properties ascribed to addition and multiplication are also basic to the understanding of certain systematic procedures for arithmetic computations. These procedures are known as *algorithms*. The following examples illustrate the relation of the fundamental laws to the familiar algorithms for addition and multiplication, respectively.

EXAMPLE 1. Find the sum $43 + 26 + 18$.

Solution: It is customary to write this in vertical form, add units to units, carry tens, and add tens to tens. Thus

$$
\begin{array}{r}
43 \\
26 \\
18 \\
\hline
87
\end{array}
$$

This procedure can be defended by looking at the problem as follows:

$$
\begin{aligned}
43 + 26 + 18 &= (40 + 3) + (20 + 6) + (10 + 8) && \text{By use of expanded notation} \\
&= (40 + 20 + 10) + (3 + 6 + 8) && \text{By commutative and associative laws} \\
&= 70 + 17 && \text{Addition of terms in parenthesis} \\
&= 70 + 10 + 7 && \text{By expanded notation} \\
&= 80 + 7 && \text{By associative law} \\
&= 87 && \text{By use of decimal numeration}
\end{aligned}
$$

What is a Number System?

EXAMPLE 2. Using the laws for natural numbers, find the product 67×4.
Solution:

$67 \times 4 = 4 \times 67$	By commutative law
$= 4 \times (60 + 7)$	By expanded notation
$= 4 \times 60 + 4 \times 7$	By distributive law
$= 240 + 28$	Multiplication of $4 \cdot 6$ tens and $4 \cdot 7$
$= 200 + 40 + 20 + 8$	By expanded notation
$= 200 + (40 + 20) + 8$	By associative law
$= 200 + 60 + 8$	Addition of terms in parenthesis
$= 268$	By use of decimal notation

In using the ordinary algorithm for this product it is customary to write

$$\begin{array}{r} 67 \\ 4 \\ \hline 268 \end{array}$$

EXERCISE 12

Peano Axioms and the Natural Numbers

If $n' = n + 1$, write the numeral that represents each of the following expressions.

1. $2'$ **2.** $17'$ **3.** $8 + 4'$
4. $(3 + 2')'$ **5.** $(5 \cdot 2)'$ **6.** $(5 \cdot 2')'$

Consider the following subsets:

$A = \{\text{even numbers}\} = \{2, 4, 6, \ldots\}$
$B = \{\text{odd numbers}\} = \{1, 3, 5, \ldots\}$
$C = \{\text{prime numbers}\} = \{2, 3, 5, 7, 11, \ldots\}$
$D = \{\text{composite numbers}\} = \{4, 6, 8, 9, \ldots\}$

Which of the following expressions is equivalent to the set N?

7. $A \cup B$ **8.** $A \cap B$ **9.** $A \cup C$
10. $C \cup D$ **11.** $\bar{D} \cup A$ **12.** $(B \cup C) \cup D$
13. $\bar{C} \cup B$ **14.** $\overline{A \cup C}$ **15.** $(B \cup \bar{C}) \cup (A \cap C)$

Using only Peano's axioms and the definitions of addition and multiplication, compute each of the following relations.

16. $3 + 1$ **17.** $1 + 3$ **18.** $3 \cdot 2$
19. $2 \cdot 3$ **20.** $2 \cdot (3 + 2)$ **21.** $2 \cdot 3 + 2 \cdot 2$

If a, b, and c are elements of the set N, which operations (if any) are represented by the symbol "$*$" in the following statements?

22. $a * b = c$ **23.** $a * b = b$ **24.** $a * b = b * a$
25. $a * (b * c) = (a * b) * c$ **26.** $a + (b * c) = (a + b) * (a + c)$

Which of the following subsets of natural numbers are closed under the given operation?

27. All even numbers under the operation of adding 6.
28. All odd numbers under the operation of adding 2.
29. All prime numbers under the operation of multiplying by 2.
30. All even numbers under the operation of multiplying by any natural number.
31. All composite numbers under the operation of adding 2.
32. All odd numbers under the operation of subtracting 2.

Using the common algorithms of arithmetic, determine each of the following.

33. What is the sum of 346, 78, and 291?
34. How much greater is 1395 than 838?
35. What is the product of 397 and 8?
36. Find the quotient of 1484 and 7.
37. $38 - (17 - 6)$ **38.** $(38 - 17) - 6$
39. $(192 \div 12) \div 2$ **40.** $192 \div (12 \div 2)$
41. $9 \cdot (18 - 3)$ **42.** $(9 \cdot 18) - (9 \cdot 3)$

43. The following example illustrates the basis of the "long multiplication" algorithm. Cite a justification for each step.

(a) $23 \cdot 34 = (20 + 3) \cdot (30 + 4)$
(b) $= (20 + 3) \cdot 30 + (20 + 3) \cdot 4$
(c) $= 30 \cdot (20 + 3) + 4 \cdot (20 + 3)$
(d) $= 600 + 90 + 80 + 12$
(e) $= 600 + 170 + 12$
(f) $= 600 + 100 + 70 + 12$
(g) $= 700 + 70 + 12$
(h) $= 782$

44. Following the procedure outlined in Problem 43, determine the product 37×28.

13. THE SYSTEM OF INTEGERS

The Number Zero

In the preceding section we discussed many of the fundamental properties of the system of natural numbers. One additional property merits our attention because it leads to the useful notion of enlarging the set of natural numbers.

In the set N there is one element b such that $a \cdot b = a$. It is called the *multiplicative identity* for the set N, and we recognize it, of course, as the number 1. It is natural to inquire whether there is a corresponding identity element for the operation of addition, that is, is there an element b such that for any $a \in N$, $a + b = a$? Since the number zero is not an element of the set N, we must conclude that there is no additive identity in the set of natural numbers.

The lack of a zero element effects a severe restriction in performing the common arithmetic operations. For this and other reasons the introduction of a new element, representing the additive identity, to our system has been hailed as one of the most important advances in mathematical history. We introduce it in the following definition.

Definition 5. The symbol "0" (read zero) represents a unique number, such that for all $a \in N$,

(1) $0 + a = a + 0 = a$ and $0 + 0 = 0$
(2) $0 \cdot a = a \cdot 0 = 0$ and $0 \cdot 0 = 0$

The definition of zero characterizes its behavior under addition and multiplication. The role of zero under the inverse operations follows directly from this definition. For example, the following three relations under subtraction result directly from part 1 of Definition 5.

(1) $a - 0 = a$
(2) $a - a = 0$
(3) $0 - 0 = 0$

These statements illustrate that in our enlarged system of numbers we can subtract zero from any number, or we can subtract any number from itself, although we cannot as yet subtract a larger number from a smaller number.

If zero is divided by any natural number the result is zero, that is,

$$\frac{0}{n} = 0 \quad \text{where} \quad n \in N$$

To verify this relation we merely observe that the product of the divisor and quotient is equal to the dividend by virtue of part 2 of Definition 5. However, division of any number *by zero* is not permitted; it has no meaning, and it is not defined. This is a result of the following considerations.

1. Assume that $n/0 = k$, where $n \neq 0$. Then it is required that $k \cdot 0 = n$. But, by the definition of zero, $k \cdot 0 = 0$. Therefore we have a contradiction.

2. Assume that $0/0 = k$. Then the definition of division as an inverse

operation requires that $k \cdot 0 = 0$. But this is true for *any* value of k. While a quotient may be said to exist, it is not unique. Therefore this division is excluded.

The Set of Whole Numbers

The enlarged set composed of the natural numbers and the number zero is called the set of *whole numbers* and will be denoted by the symbol W. Thus

$$W = N \cup \{0\} = \{0, 1, 2, 3, \ldots\}$$

The closure, commutative, associative, and distributive laws for addition and multiplication continue to hold for this enlarged set of numbers. To prove this, it is necessary only to show that these laws are true whenever the new number zero is used in combination with itself or any natural number. The closure and commutative laws follow directly from the properties of zero assigned in its definition. The remaining laws may be demonstrated as follows.

EXAMPLE 1. Verify that $(a + b) + 0 = a + (b + 0)$ where a, b, and 0 are in W.

Solution: Let $a + b = c$ since, by closure, there exists such a unique element c. Then

$c + 0 = a + (b + 0)$	By substitution of c for $a + b$ in the given equation
$c = a + (b + 0)$	By definition of zero
$ = a + b$	By definition of zero
$ = c$	By substitution of c for $a + b$

EXAMPLE 2. Verify that $a \cdot (b + 0) = a \cdot b + a \cdot 0$
Solution:

$a \cdot (b + 0) = a \cdot b + a \cdot 0$	
$a \cdot b = a \cdot b + 0$	By definition of zero
$a \cdot b = a \cdot b$	By definition of zero

The Set of Integers

By enlarging the set of natural numbers to include zero, we have devised a very serviceable system of numbers for numeration and for arithmetic operations. However, the set W is not useful for algebraic operations because even as simple a statement as $x + b = 0$, $b \neq 0$, has no replacements for x in the set W. It is therefore desirable to extend the set W by appending to it new elements which are called the *additive inverses* of the natural numbers. Hence, we assume that for each $n \in N$ there exists a corresponding number which is its additive inverse, defined as follows.

Definition 6. The additive inverse of an element n where $n \in N$ is a number denoted by $-n$, such that $n + (-n) = 0$.

The union of the set of additive inverses and the set W forms a new set whose elements are known as *integers*. If we designate this set by the symbol I, then

$$I = \{\ldots -3, -2, -1, 0, 1, 2, 3, \ldots\}$$

The numbers of the subset $\{-1, -2, -3, -4, \ldots\}$ are called *negative integers*, and the members of the subset $\{1, 2, 3, 4, \ldots\}$ are called *positive integers*. (The positive integers are sometimes written as $+1, +2, +3, +4, \ldots$.) Although zero is a member of the set of integers, it is considered neither positive nor negative.

The numbers n and $-n$ which satisfy the condition that $n + (-n) = 0$ are also said to be the opposite of each other. In this sense -5 is the opposite of 5, and 5 is the opposite of -5. As a special case the opposite of 0 is 0 since, by definition, $0 + 0 = 0$.

In more advanced courses it is proved that this enlargement of the number system can be accomplished without the introduction of any new axioms or assumptions. In such proofs each integer is defined as an ordered pair of natural numbers. The common arithmetic operations can be defined in terms of this numerical representation, and the validity of all the fundamental laws governing these operations can be confirmed wholly on the basis of the Peano axioms and the introduced definitions. Although the proof of this is left to more advanced texts, it will be instructive to examine the addition and multiplication of integers in terms of their representation as ordered pairs.

Integers as Ordered Pairs

When two numbers are paired in such a way that meaning is attached to the order in which they appear, they are called an *ordered pair*. For example, the reading 15:10 on a 24-hour clock represents an ordered pair of numbers, since the first number represents hours after midnight and the second represents minutes after the hour. Thus meaning is attached to each number by virtue of its position in the listing. Clearly the ordered pair 15:10 is not the same as 10:15. On the other hand, the prime factors of 21 do not represent an ordered pair; it makes no difference whether we list them as 3 and 7 or 7 and 3.

If a and b are natural numbers, then each integer is definable as an ordered pair represented by the notation (a, b). (Although this notation resembles that used to designate a point in graphic representation, the context in which it appears will eliminate possible confusion.) If $a > b$, the ordered pair (a, b) represents the positive integer $a - b$. For example, $(12, 5)$ represents the positive integer $12 - 5$ or 7. The numeral $(15, 8)$ is another representation for the integer 7. Similarly, $(20, 13)$, $(17, 10)$, and $(8, 1)$ each represent the positive integer 7.

If $a < b$, the ordered pair (a, b) represents the negative integer $-(b - a)$. Thus the numeral $(8, 10)$ is a representation of the negative integer -2. Other representations of this number are $(2, 4)$, $(3, 5)$, $(4, 6)$, and so on. Since each of these numerals represents the same number, we can write

$$(2, 4) = (3, 5) = (4, 6) = \cdots$$

This last statement illustrates the condition for equality of ordered pairs, which we now state as follows.

Definition 7. If (a, b) and (c, d) are ordered pairs that represent integers, then $(a, b) = (c, d)$ if and only if $a + d = b + c$.

EXAMPLE 3. Prove that $(20, 13) = (17, 10)$.
Solution: $20 + 10 = 13 + 17$. Therefore, by Definition 7, $(20, 13) = (17, 10)$.

EXAMPLE 4. For what value of b will the ordered pairs $(2, 5)$ and $(8, b)$ be equal?
Solution: $(2, 5) = (8, b)$ if and only if $2 + b = 5 + 8$
Therefore
$$2 + b = 13$$
$$b = 11$$

Operations with Integers

Addition, of course, can be defined arbitrarily in the system of integers. However, since the set of integers is an extension of the set of natural numbers, we are motivated to frame our definition in such a way as to preserve the addition relations for natural numbers. For example, the addition relation $5 + 3 = 8$ can be written in terms of ordered pairs as

$$(7, 2) + (4, 1) = (11, 3)$$

We note that the component 11 is the sum of the first components 7 and 4 and the component 3 is the sum of the last components 2 and 1. This consideration is the basis of the following definition.

Definition 8. For any two integers (a, b) and (c, d),
$$(a, b) + (c, d) = (a + c, b + d)$$
Applying this definition to various combinations of positive and negative integers, as illustrated in the following examples, we obtain what should be familiar results:

$(8, 4) + (5, 2) = (13, 6),$ that is, $(+4) + (+3) = +7$
$(4, 6) + (1, 7) = (5, 13),$ that is, $(-2) + (-6) = -8$
$(7, 2) + (3, 12) = (10, 14),$ that is, $(+5) + (-9) = -4$
$(3, 9) + (3, 2) = (6, 11),$ that is, $(-6) + (+1) = -5$

These results can also be written in vertical form, as follows:

$$
\begin{array}{cccc}
+4 & -2 & +5 & -6 \\
+3 & -6 & -9 & +1 \\
\hline
+7 & -8 & -4 & -5
\end{array}
$$

Generalizing from these examples, we may now formulate the following familiar rules for the addition of integers.

1. In adding integers with like signs, find the sum of their magnitudes and prefix the common sign.
2. In adding integers with unlike signs, find the difference of their magnitudes and prefix the sign of the larger magnitude.

It is to be expected, of course, that the multiplication of integers can also be developed in terms of ordered pairs. The following definition introduces this concept.

Definition 9. For any two integers (a, b) and (c, d),
$$(a, b) \cdot (c, d) = (ac + bd, bc + ad)$$
As examples of the application of this definition to various combinations of positive and negative integers, we have:

$(5, 2) \cdot (7, 3) = (35 + 6, 14 + 15) = (41, 29)$ or $(+3) \cdot (+4) = +12$
$(1, 3) \cdot (2, 7) = (2 + 21, 6 + 7) = (23, 13)$ or $(-2) \cdot (-5) = +10$
$(6, 2) \cdot (3, 9) = (18 + 18, 6 + 54) = (36, 60)$ or $(+4) \cdot (-6) = -24$
$(1, 4) \cdot (10, 3) = (10 + 12, 40 + 3) = (22, 43)$ or $(-3) \cdot (+7) = -21$

Thus the usual rule of signs for multiplication of integers evolves. It may be stated as follows.

> The sign associated with the product of two integers having like signs is positive; the sign associated with the product of two integers having unlike signs is negative.

13. The System of Integers

Procedures for subtraction and division of members of the set of integers can now be developed as a necessary consequence of their role in inverse operations. For example, to determine the result of the subtraction $(+3) - (+7) = \Box$, we are really asking what integer added to $+7$ will produce $+3$. We know that $(+7) + (-4) = +3$, therefore, $(+3) - (+7) = -4$. In a similar manner, we can show that $(+2) - (-9) = 11, (-5) - (-4) = -1$, and $(-6) - (+8) = -14$. These results and any others we may wish to investigate, lead to the following rule for the subtraction of integers.

> In subtracting one integer from another, change the sign of the subtrahend mentally and proceed as in addition.

The analogous situation with respect to division of integers is based on a consideration of the following types of problems:

$$\frac{+12}{+2} = \Box \qquad \frac{+21}{-3} = \Box \qquad \frac{-15}{+5} = \Box \qquad \frac{-36}{-9} = \Box$$

In each case the quotient must satisfy a corresponding multiplication relation. Thus, the first quotient is 6 since $6 \cdot (+2) = +12$, the second quotient is -7 since $(-7) \cdot (-3) = +21$, and the last two quotients are -3 and 4, respectively. Therefore, we note that the rule for the sign of the quotient of two integers follows the same form as that for multiplication, that is:

> The sign associated with the quotient of two integers is positive if the two integers have like signs, and it is negative if they have unlike signs.

EXERCISE 13

THE SYSTEM OF INTEGERS

Which of the following operations with zero have no meaning? Where the operation is possible, indicate the result.

1. $0 \div 8$
2. $0 + 5$
3. $7 \cdot 0$
4. $0 - 7$
5. $0 \div 0$
6. $0 + 0$
7. $11 \div 0$
8. $0 \div (-3)$
9. $-4 \cdot 0$

Using the definition of zero, prove each of the following relations for the set of whole numbers.

10. $(a + 0) + b = a + (0 + b)$
11. $(0 + a) + b = 0 + (a + b)$
12. $a \cdot (0 + b) = a \cdot 0 + a \cdot b$
13. $0 \cdot (a + b) = 0 \cdot a + 0 \cdot b$

What is a Number System?

Write the integer represented by each of the following ordered pairs.

14. (8, 3) **15.** (4, 2) **16.** (11, 1) **17.** (4, 0)
18. (0, 8) **19.** (2, 6) **20.** (5, 6) **21.** (3, 3)
22. (15, 4) **23.** (2, 19) **24.** (1, 13) **25.** (9, 9)

Determine whether each of the following is or is not equal to the ordered pair (5, 8). Prove your response by referring to the definition of equality of ordered pairs.

26. (0, 3) **27.** (1, 5) **28.** (8, 5)
29. (4, 7) **30.** (5, 2) **31.** (10, 16)

Express the result of each of the following operations as an ordered pair.

32. (3, 1) + (5, 7) **33.** (4, 6) + (5, 9) **34.** (2, 5) · (3, 1)
35. (4, 4) · (8, 2) **36.** (7, 8) + (2, 2) **37.** (1, 7) + (7, 1)
38. (8, 3) · (4, 5) **39.** (2, 3) · (3, 2) **40.** (4, 9) · (1, 6)

Replace the symbol □ with an element of the set of integers in such a way as to make each of the following statements true.

41. 3 + □ = 5 **42.** 5 · □ = −30 **43.** □ + 6 = 1
44. 18 ÷ □ = −9 **45.** −2 · □ = 10 **46.** 7 − □ = −2
47. □ · 3 = −33 **48.** □ − (−2) = 17 **49.** □ ÷ 8 = −4

Determine the value of $a + b$, $a \cdot b$, $a - b$, and $a \div b$ for the following values of a and b respectively.

50. 9 and 3 **51.** 15 and 5 **52.** 8 and −2 **53.** 12 and −4
54. −10 and 5 **55.** −18 and 6 **56.** −5 and −1 **57.** −16 and −8
58. 6 and −6 **59.** −3 and 3 **60.** 7 and 7 **61.** −4 and −4
62. 0 and −2 **63.** 0 and 6 **64.** −12 and 3 **65.** −8 and −4

If a and b are natural numbers, then the ordered pair $(a, 0)$ represents a positive integer, the ordered pair $(0, b)$ represents a negative integer, and the ordered pair (a, a) represents zero. Using this form of notation show that:

66. The additive inverse of $(a, 0)$ is a negative integer.
67. The additive inverse of $(0, b)$ is a positive integer.
68. The sum of two positive integers is a positive integer.
69. The sum of two negative integers is a negative integer.
70. The product of two positive integers is a positive integer.
71. The product of two negative integers is a positive integer.
72. The product of a positive integer and a negative integer is a negative integer.

14. THE RATIONAL NUMBERS

In the last section the integers were introduced as *differences* of natural numbers. We will now introduce new numbers which are *quotients* of integers. Our objective is to construct a number system of greater versatility, one that will contain a multiplicative inverse as well as an additive inverse for each element in the set. The need for such an enlargement of our number system is apparent from a consideration of the highly restrictive conditions under which division is possible within the set I, the set of integers.

If we wish to apportion 16 like objects among n people, the apportionment is possible within the set I only if n is 1, 2, 4, 8, or 16. There is no element in the set I for the indicated division 16/3, or 16/5, or 16/6. If division is to be made possible for all integers, we must create an expanded number system which will contain a new number x such that $x = a/b$ (or the equivalent relation $b \cdot x = a$) for all admissible values of a and b in I. The new number x which is designed to meet this condition is called a *rational number*.

Rational Numbers as Ordered Pairs

Just as we had used the ordered pair (a, b) to denote the integer x such that $x + b = a$, let us denote here by (a, b) the number x such that $x \cdot b = a$. For example, the rational number that satisfies the condition $3x = 7$ is $(7, 3)$, and the rational number that satisfies the condition $5x = 2$ is $(2, 5)$. In the form of fractions, of course, these are customarily written as 7/3 and 2/5, respectively. We can see that we must add a restriction to such ordered pairs (a, b), namely, $b \neq 0$. Since the introduction of rational numbers establishes the existance of a number x which always satisfies the general multiplication relation $bx = a$, $b \neq 0$, we have, in effect, removed all restrictions against the operation of division, that is, for all a and b in I, $a \div b = (a, b)$, $b \neq 0$. We now extend the role of these new numbers within the system by means of the following definitions, given the two rational numbers (a, b) and (c, d).

Definition 10. Equality: $(a, b) = (c, d)$ if and only if $ad = bc$.
Definition 11. Addition: $(a, b) + (c, d) = (ad + bc, bd)$.
Definition 12. Multiplication: $(a, b) \cdot (c, d) = (ac, bd)$.

EXAMPLE 1. Show that the rational number that satisfies the relation $3x = 7$ is equal to the number that satisfies the relation $6y = 14$.

Solution: By definition of rational numbers as ordered pairs,
$$x = (7, 3)$$
$$y = (14, 6)$$
By Definition 10, $(7, 3) = (14, 6)$ because $7 \cdot 6 = 14 \cdot 3$. Therefore, $x = y$.

EXAMPLE 2. Express the sum of the two rational numbers $(3, 5)$ and $(1, 2)$ as an ordered pair.
Solution: By Definition 11, $(3, 5) + (1, 2) = (6 + 5, 10) = (11, 10)$.

EXAMPLE 3. Express the product $(2, 7) \cdot (4, 3)$ as an ordered pair.
Solution: By Definition 12, $(2, 7) \cdot (4, 3) = (8, 21)$.

Rational Numbers as Fractions

The notation (a, b) has been used to represent rational numbers in order to demonstrate the importance of definitions in developing the system of rational numbers. A more common notation as well as a more useful one for purposes of calculation is the familiar symbol $\frac{a}{b}$ or a/b. Expressed in this form, the rational number is called a *fraction* and the numbers a and b of the fraction a/b are called the *numerator* and *denominator*, respectively. The fraction a/b is subject to the following interpretations.

1. *As a partition.* The fraction 2/5 means that some quantity is partitioned into 5 equal parts, and we are concerned with 2 of those parts. The quantity under consideration may be a length, an area, a collection of objects, or any other measureable or countable entity.

2. *As a division.* This meaning is not inconsistent with the interpretation of a fraction as a partition; it merely gives us additional ability to apply these numbers to certain situations. For example, 2/5 means $2 \div 5$. If an entity of 2 units, say, 2 dollars, is divided into 5 equal parts, then each part is 2/5 of a dollar. The same result, of course, is obtained as a partition; 2 of the 5 equal parts of one dollar is 2/5 of a dollar.

3. *As a ratio.* This meaning of a fraction permits still broader applications, without contradicting either of the previous interpretations. It is concerned with the comparison of two measures. Ordinarily we can express an *absolute comparison* of the two measures by stating their difference, or we can express a *relative comparison* of two measures by stating their quotient. The latter comparison is called a *ratio*. When we say that 6 hours is 4 more than 2 hours, we are expressing an absolute comparison, but when we say that 6 hours is 3 times as great as 2 hours, or that the ratio of 6 hours to 2 hours is 3 to 1, we are expressing a relative comparison.

Having reviewed the notion of a fraction and its interpretations, we now define the conditions under which two of these symbols represent the same number. Since this has already been done for the representation of these numbers as ordered pairs, we need merely to express the definition in terms of fractions, as follows.

Definition 13. $a/b = c/d$ if and only if $ad = bc$.

This definition is the basis of the following useful relation:

$$\frac{a}{b} = \frac{ka}{kb}, \qquad k \neq 0$$

which can be proved by observing that $a(kb) = b(ka)$, since the associative and commutative properties of integers are applicable. One direct consequence of this relation is that we can find as many equivalent forms for a given rational number as we like. For example,

$$\frac{3}{7} = \frac{6}{14} = \frac{9}{21} = \frac{12}{28} = \cdots$$

Another useful consequence of this relation is to effect a "reduction" in the form of the fraction by identifying a common factor k in the terms of the fraction, as illustrated in the following examples:

$$\frac{33}{39} = \frac{11 \times 3}{13 \times 3} = \frac{11}{13}$$

$$\frac{14}{70} = \frac{14 \times 1}{14 \times 5} = \frac{1}{5}$$

The procedures for the fundamental operations with fractions now follow directly from the definitions listed for the corresponding operations with ordered pairs. The sum of any two rational numbers expressed as fractions is defined by the relation

$$\frac{a}{b} + \frac{c}{d} = \frac{ad + bc}{bd}$$

and the product of two such numbers is defined by the relation

$$\frac{a}{b} \cdot \frac{c}{d} = \frac{ac}{bd}$$

Thus,
$$\frac{2}{3} + \frac{5}{7} = \frac{14 + 15}{21} = \frac{29}{21}$$

and
$$\frac{2}{3} \cdot \frac{5}{7} = \frac{2 \times 5}{3 \times 7} = \frac{10}{21}$$

A negative rational number may be expressed as $-(a/b)$, $-a/b$, or $a/-b$, where a and b are natural numbers. Any one of these equivalent forms is the additive inverse of the rational number a/b, that is, $a/b + (-a/b) = 0$, by direct application of the definition of addition. Therefore, as in the case of integers, subtraction is definable in terms of an equivalent addition. For example,

$$\frac{7}{5} - \frac{1}{2} = \frac{7}{5} + \frac{-1}{2} = \frac{14 + (-5)}{10} = \frac{9}{10}$$

The extension of the number system to include rationals was motivated by a need for a *multiplicative inverse* corresponding to each element in the set except zero. We now define this inverse in terms of fractional notation and apply it to the development of a division algorithm for fractions.

Definition 14. The multiplicative inverse of an element a/b, where a/b is a rational number, is a number x such that $a/b \cdot x = 1$.

Thus, the multiplicative inverse of 3/7 is 7/3, the multiplicative inverse of 5 is 1/5 and, in general, the multiplicative inverse of a/b is b/a. The numbers a/b and b/a are also frequently referred to as *reciprocal numbers*.

Now consider the indicated division $a/b \div c/d$, $c/d \neq 0$, which we can write as

$$\frac{\frac{a}{b}}{\frac{c}{d}} = \frac{\frac{a}{b} \cdot \frac{d}{c}}{\frac{c}{d} \cdot \frac{d}{c}} = \frac{\frac{a}{b} \cdot \frac{d}{c}}{1}$$

Therefore, $$\frac{a}{b} \div \frac{c}{d} = \frac{a}{b} \cdot \frac{d}{c}$$

Thus we arrive at the familiar algorithm: to divide by a nonzero number, multiply by its reciprocal.

The Set of Rational Numbers

In summary, a rational number is one that can be expressed as a quotient of two integers, with the restriction that the denominator cannot be zero. The set of rationals, designated by the symbol R, is an extension of the set of integers and contains the integers as a subset, that is, $I \subset R$.

The rational numbers are closed, associative, and commutative with respect to both addition and multiplication. Further, the set R contains an identity element 0 for addition, an identity element 1 for multiplication, an additive inverse $-n$ for each element n in the set, and a multiplicative inverse b/a for each nonzero element a/b in the set. Finally, the multiplication of rational numbers is distributive with respect to addition.

Thus, the set R contains the natural numbers as a subset but overcomes their lack of an additive identity; it contains the whole numbers as a subset but overcomes their lack of additive inverses; and it contains the integers as a subset but overcomes their lack of multiplicative inverses. In this sense, the rationals represent the most complete, most useful, and most versatile set of numbers we have considered up to this point.

EXERCISE 14

THE RATIONAL NUMBERS

Determine which of the following ordered pairs represent equal rational numbers.

1. (3, 5), (5, 3)
2. (7, 3), (6, 14)
3. (2, 5), (10, 25)
4. (4, 1), (12, 3)
5. (−2, 3), (2, −3)
6. (7, −4), (−14, 8)

Using the definitions for addition and multiplication of rational numbers, express the result of each indicated operation as an ordered pair.

7. (2, 3) + (3, 2)
8. (7, 5) + (2, 1)
9. (3, 5) · (2, 7)
10. (6, 5) · (11, 1)
11. (−2, 5) + (1, 9)
12. (3, −4) + (−2, 5)
13. (7, 6) + (−5, 7)
14. (−2, 3) · (4, 11)
15. (−3, 8) · (−3, 4)

Verify each of the following relations for rational numbers.

16. (4, 5) + (2, 3) = (2, 3) + (4, 5)
17. (7, 3) · (2, 5) = (2, 5) · (7, 3)
18. (−2, 9) · (−1, 5) = (−1, 5) · (−2, 9)
19. (−4, 13) + (−1, 2) = (−1, 2) + (−4, 13)
20. (1, 3) + [(2, 5) + (3, 7)] = [(1, 3) + (2, 5)] + (3, 7)
21. (3, 2) · [(1, 2) + (5, 7)] = (3, 2) · (1, 2) + (3, 2) · (5, 7)

State, in words, three possible interpretations of each of the following fractions.

22. 3/5
23. 8/3
24. 14/7
25. 2/9

Write four indicated quotients representing each rational number.

26. 1/3
27. 2/7
28. 5/2
29. 9/8

Find in simplest form the rational number that represents each of the following expressions.

30. 2/3 + 1/5
31. 5/2 + 1/4
32. 3/7 − 2/5
33. 4/3 − 5/8
34. 7/3 · 2/9
35. 2/5 · 2/4
36. 1/8 ÷ 2/3
37. 5/7 ÷ 2/9
38. 9/4 + (−6)/5

Write (a) the additive inverse and (b) the multiplicative inverse of each of the following rational numbers.

39. (4, 5)   **40.** (−2, 7)   **41.** (8, 1)   **42.** (14, −3)
43. 3/4   **44.** 5/8   **45.** −8/3   **46.** 6

47. By reference to the definition of the addition of fractions, prove that

$$\frac{a}{c} + \frac{b}{c} = \frac{a+b}{c}$$

Using the result of Problem 47, express each of the following sums as a quotient of two integers.

48. 5/7 + 1/7   **49.** 3/8 + 4/8   **50.** 2/9 + 5/9

Convert each of the following fractions into equivalent forms with like denominators, and complete the computation using the relation stated in Problem 47.

51. 1/2 + 1/4   **52.** 2/3 + 1/6   **53.** 3/5 + 2/15
54. 3/2 + 1/3   **55.** 2/5 + 4/7   **56.** 5/8 + 1/12

Add each of the following by (a) repeated application of the definition of addition of fractions and (b) using the common-denominator approach.

57. 1/3 + 2/5 + 5/6   **58.** 2/7 + 1/2 + 4/9
59. 2/7 + 1/3 + 3/5 + 3/10   **60.** 1/4 + 2/3 + 2/5 + 5/12

61. Prove that $a/b - c/d$ is a rational number, that is, that the set of rationals is closed under subtraction.
62. Prove that $a/b \div c/d$, $c/d \neq 0$, is a rational number, that is, that the set of rationals is closed under division.
63. Prove that the commutative law holds for the addition of rational numbers.
64. Prove that the distributive law holds for multiplication with respect to addition for rational numbers.

15. THE REAL NUMBER SYSTEM

It might appear at this point that the system of rational numbers is a final and complete number system in that no further extension is required. From the standpoint of the operations of addition, multiplication, and their inverses, this is so. However, other operations as well as geometric considerations dictate a need for recognizing numbers that fall outside the set of rationals.

A number that cannot be expressed as a quotient of two integers is called an *irrational number*. The possibility of the existence of such numbers is not obvious, and the need to recognize them was very disquieting to early mathematicians. Although a definition of irrational numbers based on successive ratios was developed as early as the 4th century B.C., it was not until the 19th century that all ambiguities and reservations concerning these numbers were removed. Their role in a rigorous and logical mathematical structure was established principally through the efforts of such relatively modern mathematicians as Richard Dedekind and Karl Weierstrass.

All this indicates that the concept of irrationality of numbers is not one which is intuitively simple, even though these numbers are used today as comfortably and as rigorously as the rationals. There are various ways of introducing irrational numbers; the method we shall use has the advantage of describing them in a familiar form.

Decimal Notation

It is assumed that the reader is familiar with the ordinary uses of decimal notation. If a rational number is expressed as a quotient of two integers, we can determine a decimal representation for the rational number by simply performing the indicated division. One of two situations will always prevail.
 1. The decimal representation will *terminate*, as in $3/8 = 0.375$, or
 2. The decimal representation will *repeat*, as in $3/11 = 0.2727\ldots$.

The proof that the decimal representation of a rational number will either terminate or repeat is not difficult. It is evident that in the division process only a finite number of different remainders can occur. If one of these remainders is zero, the decimal terminates; if all remainders are nonzero, then one of them must be repeated, and ultimately this repetition must occur after the stage at which zeros are annexed to the dividend. The sequence of digits in the quotient will then continue in a repeating cycle.

EXAMPLE 1. Express the rational number 7/12 in decimal notation.
Solution: $7/12 = 7 \div 12 = 0.58333\ldots$.

EXAMPLE 2. What is the fractional equivalent of the terminating decimal 2.03?
Solution: $2.03 = 2/1 + 3/100 = 203/100$.

EXAMPLE 3. Show that the repeating decimal $4.222\cdots$ is a rational number.
Solution: (1)   Let $n = 4.222\ldots$
 (2) Then $10n = 42.222\ldots$
 (3) $9n = 38$ By subtraction of (1) from (2)
 (4) $n = 38/9$ By definition of division

These examples illustrate that every rational number can be written as either a terminating or a repeating decimal and, conversely, every terminating or repeating decimal can be written as a quotient of two integers. It follows that an irrational number, when expressed in decimal notation, will neither terminate nor develop a repeating cycle of digits. We will introduce this concept by examining a common example of an irrational number—the square root of 2.

The square root of a number is defined to be one of its two equal factors. It is the inverse of the operation of finding the square of a number, that is, determining the product of the number multiplied by itself. Thus, the square root of 16 is 4 because $4 \cdot 4 = 16$, and the square root of 16 is also -4 because $(-4) \cdot (-4) = 16$. Now the square root of 2, represented by the symbol $\sqrt{2}$, is a number x such that $x \cdot x = 2$ or $2 \div x = x$. That this number is irrational is certainly not obvious from a cursory inspection or by random trial of replacements from the set of rational numbers. One way to examine its decimal representation is to set up a division process that approximates the value x in the relation $2 \div x = x$. The process that follows is an interesting and useful procedure—interesting because the further it is carried out, the closer we approach the value x, and useful because it is a repetitious process which is easily adaptable to programming on modern computers. For this reason, it is called an *iterative* process (from the word "iterate" which means "to repeat").

Iterative Process for Approximating $\sqrt{2}$[1]

Step 1. Estimate $\sqrt{2} = 1.3$.
Step 2. Divide $2 \div 1.3 = 1.54$.
Step 3. Average $\dfrac{1.3 + 1.54}{2} = 1.41$.

Repeat Steps 2 and 3: $\qquad 2 \div 1.41 = 1.418$

$\dfrac{1.41 + 1.418}{2} = 1.414$

$2 \div 1.414 = 1.414428$

$\dfrac{1.414 + 1.414428}{2} = 1.414214$

$2 \div 1.414214 = 1.414213$

---

[1] It is not difficult to see why the process works. We are seeking a value x such that $2 \div x = x$. If we choose a value $x_1 < x$ as a first estimate, then $2 \div x_1 = x_2$ where $x_2 > x$. Hence x is between x_1 and x_2, and taking the average is, therefore, a plausible way of selecting a value of x between x_1 and x_2.

Our computations show that the decimal representation of $\sqrt{2}$, correct to five decimal places, is 1.41421 and, further, that $1.414213 < \sqrt{2} < 1.414214$. It is also evident that the process can be carried forward as far as we please, but it is not evident that the process will fail to terminate or to produce a repeating decimal no matter how long or diligently we work. How, then, can we be certain that $\sqrt{2}$ is not a rational number? The proof is an interesting example of mathematical reasoning.

Proof of Irrationality of $\sqrt{2}$.

The proof of the irrationality of $\sqrt{2}$ is based on the understanding of a simple property of integers that are perfect squares, such as 9, 36, and 400. If we write each number in terms of its prime factors, then

$$9 = 3 \cdot 3$$
$$36 = 6 \cdot 6 = 2 \cdot 3 \cdot 2 \cdot 3$$
$$400 = 20 \cdot 20 = 2 \cdot 2 \cdot 5 \cdot 2 \cdot 2 \cdot 5$$

We observe that an integer is a perfect square if and only if each of its prime factors occurs an even number of times. It is evident that this property is not restricted to the specific numbers 9, 36, or 400. In general, if an integer is a perfect square it has factors of the form $r \cdot r$. If r is prime, then r occurs an even number of times; if r is not prime, then each prime factor of r will reoccur in the second factor and, therefore, must occur an even number of times. Conversely, if prime factors occur an even number of times, their product will be, by definition, the square of an integer. We now return to the problem of proving that no rational number exists for the representation of $\sqrt{2}$.

The central idea in the proof is to *assume* that $\sqrt{2}$ is a rational number and then show that this leads to an impossibility. Therefore we assume that

$$\sqrt{2} = \frac{a}{b}$$

where a and b are integers that have no common factor. Then

$\quad \dfrac{a}{b} \cdot \dfrac{a}{b} = 2 \qquad$ By definition of square root

$\quad \dfrac{a^2}{b^2} = 2 \qquad$ By definition of multiplication of rationals

$\quad a^2 = 2b^2 \qquad$ By definition of inverse operations

Now let us examine this last statement very carefully. The number a^2 is the square of an integer, therefore, 2 is a factor of a^2 either an even number of times or not at all. Similarly, 2 is a factor of b^2 either an even number of times or not at all. In either case, the expression $2b^2$ must have a factor 2 an odd number of times. But this is impossible since the numbers a^2 and $2b^2$ are equal. Since we have reached an impossible conclusion, our original assumption, that $\sqrt{2}$ is rational, cannot be sustained. Therefore, $\sqrt{2}$ is an irrational number.

The Set of Irrational Numbers

Having established that no rational number corresponds to the square root of 2, a natural line of inquiry is to investigate the existence of other irrational numbers. A little reflection will show that the proof used for $\sqrt{2}$ can be used to show that $\sqrt{3}$ is irrational or that the square root of any prime number is irrational. For example, to show that $\sqrt{5}$ is irrational, we merely begin with the assumption $\sqrt{5} = a/b$. This results in the statement $a^2 = 5b^2$. It is easily verified that 5 must appear as a prime factor an odd number of times in the expression $5b^2$, and an even number of times in the expression a^2. Therefore, we again reach a contradiction which establishes that $\sqrt{5}$ cannot be rational, hence, it is irrational. Further, by only a slight modification in the procedure it can be shown that the only integers that have rational square roots are those which are perfect squares of integers. Therefore, each of the following is irrational:

$$\sqrt{2}, \sqrt{3}, \sqrt{5}, \sqrt{6}, \sqrt{7}, \sqrt{8}, \sqrt{10}, \sqrt{11}, \sqrt{12}, \sqrt{13}, \ldots$$

Similar proofs can be developed to show that the only integers that have rational cube roots (one of three equal factors) are those which are perfect cubes of integers. Thus, the cube root of each of the numbers 2, 3, 4, 5, 6, or 7 is irrational, but the cube root of 8 is rational.

It is also possible to contrive decimal representations of irrational numbers. For example,

$$3.21211211121111 \cdots$$

is an irrational number if the pattern of development is continued, since the decimal will neither terminate nor repeat. But such contrived numbers are of only theoretical interest and represent a trivial illustration of irrational numbers. Of far greater importance are irrationals which occur naturally in mathematics and science. A good example of such a number in the field of elementary mathematics is the number π, which is the ratio of the circumference of a circle to its diameter. Many other irrationals, however, will be encountered in a further study of mathematics.

15. The Real Number System

Finally, additional irrationals may be generated by combining them with rational numbers through the use of the fundamental operations. The approach to the investigation of such expressions is illustrated in the following examples.

EXAMPLE 4. Prove that $2 + \sqrt{5}$ is irrational.
Solution: Assume that $2 + \sqrt{5} = r$, where r is rational. Then $\sqrt{5} = r - 2$, by definition of inverse operations. Therefore, $\sqrt{5}$ is rational because the difference of two rational numbers is rational. But $\sqrt{5}$ has been proved irrational. Since our assumption has led to a contradiction, the alternative conclusion that $2 + \sqrt{5}$ is irrational is established.

EXAMPLE 5. Prove that if n is a rational number other than zero, then $d\sqrt{2}$ is irrational.
Solution: Assume that $n\sqrt{2} = k$, where k is a rational number. Then $\sqrt{2} = k/n$, by definition of inverse operations. Therefore, $\sqrt{2}$ is a rational number since k and n are rationals. But $\sqrt{2}$ has been proved irrational. Therefore our assumption that $n\sqrt{2}$ is a rational number is untenable; we must conclude it is irrational.

The Real Numbers

If we designate the set of rationals as R, and the set of irrationals as $\bar{R}$, then the union of these two sets constitutes the set Z, the set of real numbers. Thus $Z = R \cup \bar{R}$. The relationship of the set of real numbers to the other sets we have considered is shown in Figure 2.7.

A further insight into the relation of these number sets may be obtained from a graphic representation which associates each number with a point on

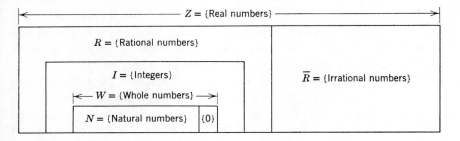

Figure 2.7 Subsets of the real numbers.

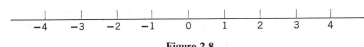

Figure 2.8

a line. The straight line shown in Figure 2.8 is considered to extend indefinitely in both directions. At some arbitrary point on the line, we have designated a starting position or point corresponding to the number zero. We call this point the *origin*. Starting with the origin we have marked off equal line segments, associating with each successive point to the right of the origin the natural numbers 1, 2, 3, ..., and with each successive point to the left of the origin the negative integers $-1, -2, -3, \ldots$, The integers on the number line are equally spaced, and it is a simple matter to identify the integer that follows any given integer. This property does not apply to the set of rationals; we cannot identify the rational number that is "next larger" than a given rational number. For example, given the number 1/3, what is the next larger rational number? If we divide the segment from 0 to 1 into three equal parts, the points of division would correspond to the rational numbers 0, 1/3, 2/3, and 3/3, respectively. Then 2/3 appears to be the rational number that follows 1/3. But if we divide the same segment into six equal parts, then the points of division correspond to the rational numbers 0, 1/6, 2/6, 3/6, 4/6, 5/6, and 6/6, respectively. Now 3/6 (or 1/2) appears to be the rational number that follows 2/6 (or 1/3). Clearly we could continue subdividing the given line segment into smaller and smaller parts, with the result that the points of division would be closer and closer to each other. There is no limit to the closeness of the resulting points and, therefore, we can never determine a number that is "the next larger rational number" to the given number.

In the process it may appear that every point on the line must ultimately be associated with some rational number. After all, as we draw in more and more division marks, it is hard to imagine that any point could escape being a mark corresponding to some point of division. However, our experience with irrational numbers tells us that this is so. For example, the point corresponding to $\sqrt{2}$ cannot be located by subdividing the segment between 1 and 2, no matter how refined our subdivision may be. That a point corresponding to $\sqrt{2}$ does exist is apparent from a consideration of Figure 2.9.

By a simple application of the Pythagorean Theorem, the right triangle, whose legs are each 1 unit long, has a hypotenuse whose length is $\sqrt{2}$. Therefore, if such a triangle is constructed on the number line, as illustrated in Figure 2.9, the terminal point of the hypotenuse will determine a point that corresponds to the irrational number $\sqrt{2}$. Similarly, each irrational number can be associated with a unique point on the number line.

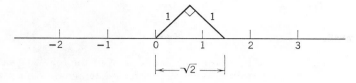

Figure 2.9

The ultimate result is that each point on the number line is associated with one and only one rational or irrational number and, conversely, each such number is associated with one and only one point. Thus there is a one-to-one correspondence between the points on the number line and the elements of the set of real numbers.

We have noted throughout these discussions the important properties of number systems—properties relating to closure, commutativity, associativity, and distributivity. Of course, the extension of the number system from rationals to real numbers does not routinely preserve all the properties developed for the rationals, but fortunately this proves to be so. However, to demonstrate this in detail would be tedious and complicated; our interest at this stage is limited to suggesting the motivation for extending a number system and introducing some of the mathematical ideas involved in such extensions. Although still further enlargements of the number system are both possible and useful, we will agree that hereafter in this book, unless otherwise specified, the word number will mean real number.

EXERCISE 15

THE REAL NUMBER SYSTEM

If Z is the set of real numbers, R the set of rationals, $\bar{R}$ the set of irrationals, I the set of integers, W the set of whole numbers, and N the set of natural numbers, list *all* sets to which each of the following numbers belongs.

1. 4
2. -7
3. 2/3
4. $-4/5$
5. $\sqrt{10}$
6. 0.75
7. 0.7575 $\cdots$
8. 0
9. $\sqrt{49}$

Express each of the following as a decimal.

10. 3/8
11. 5/12
12. 4/13
13. 7/9
14. 6/7
15. 7/6

Express each of the following as a quotient of two integers.

16. 0.24
17. 0.13

102 *What is a Number System?*

18. $0.2121 \cdots$ (Let $n = 0.2121 \cdots$. Then $100n = ?$)
19. $4.777 \cdots$ **20.** 6.03
21. $5.871871 \cdots$

Indicate whether the decimal representation of each of the following numbers will (a) terminate, (b) not terminate but repeat, or (c) not terminate and not repeat.

22. $\sqrt{6}$ **23.** $\sqrt{8}$ **24.** $3/5$
25. $5/3$ **26.** $\sqrt{0.01}$ **27.** $\sqrt{25/36}$

Using the iterative process, approximate the decimal reprseentation of each of the following numbers to three digits.

28. $\sqrt{7.43}$ **29.** $\sqrt{1.52}$ **30.** $\sqrt{0.781}$
31. $\sqrt{13.3}$ **32.** $\sqrt{133}$ **33.** $\sqrt{1330}$

34. Prove that $\sqrt{3}$ is an irrational number by a method similar to that used for $\sqrt{2}$.
35. Prove that $\sqrt{7}$ is not rational.
36. Prove that if $\sqrt{3}$ is irrational, then $2\sqrt{3}$ is irrational.
37. Prove that if $\sqrt{5}$ is irrational, then $k + \sqrt{5}$ is irrational, where k is any rational number.
38. Show by an example that if m and n are real numbers different from zero, then (a) $\sqrt{m} + \sqrt{n} \neq \sqrt{m+n}$ and (b) $\sqrt{m} \cdot \sqrt{n} = \sqrt{m \cdot n}$.
39. Given the two rational numbers $a = 2.3131 \cdots$ and $b = 2.3232 \cdots$, and the rational number $c = 2.317$, which lies between them on the number line. (a) List two more rational numbers which lie between a and b. (b) How many other rational numbers exist between a and b? (c) If $d = 2.314315316 \cdots$, then d is an irrational number that lies between a and b. List two more irrational numbers that lie between a and b. (d) How many other irrational numbers exist between a and b?
40. Given to four decimal places the two irrational numbers $p = \sqrt{2} = 1.4142$ and $q = \sqrt{2.01} = 1.4177$. (a) List two rational numbers that lie between p and q on the number line. (b) List two irrational numbers that fall between p and q. (c) How many other rational numbers exist between the irrationals p and q? (d) How many other irrational numbers exist between p and q?

3

What is the Axiomatic Basis of Algebra?

16. THE EQUALITY RELATION

In the early 19th century, algebra was considered simply as a generalization of arithmetic, and this is still an influential view in much of the teaching of elementary algebra. Considered in this way, algebra uses letters as symbols to represent numbers and develops useful procedures for manipulations of these symbols in fundamental operations, equations, inequalities, and problem solving.

A more modern view of algebra emphasizes the *structure* of algebra. This view is based on the concept that, since the statements of algebra are symbolic, they can be applied not only to numbers but to other kinds of elements as well. For example, algebraic statements can be applied to sets of points in a plane, with addition defined as the union of two sets and multiplication defined as the intersection of two sets. Similarly, algebraic statements can be applied to number pairs, geometric configurations, or permutations of numbers, providing that in each case appropriate definitions for the basic operations of addition and multiplication are prescribed.

Since algebra can, therefore, be severed from its tie with arithmetic, it is important to examine the premises on which it is based. We find that algebra

104   *What is the Axiomatic Basis of Algebra?*

makes use of axioms in much the same way that, traditionally, geometry has done. Although a complete development of elementary algebra from axioms would be too abstract for our purposes, an introduction to some of the basic premises of algebra is essential to any understanding that proceeds beyond a superficial stage.

Relations

Every student is no doubt familiar with the extensive use of the equality sign ($=$) in algebraic discourse. Equality is a *relation*, with certain properties which can be stated in the form of axioms. The concept of relation occurs frequently both within and outside the field of mathematics. It is implicit in phrases such as:

is married to	is equal to
is wealthier than	is the employer of
is congruent to	is parallel to
is less than	is at least as large as
is a representative of	is a subset of

Although relations are always concerned with the connectivity of two or more elements, it is possible to consider the general properties of relations independent of the particular elements under discussion. In a consideration of such properties, it will be useful to express symbolically the fact that two elements are connected by a given relation. If x is related to y through the relation R, we shall write xRy, which is read "x has the relation R to y." We shall now discuss the following properties as they apply to a relation between two and only two elements:

Reflexiveness
Symmetry
Transitivity

A relation R is said to be *reflexive* whenever xRx. For example, congruence is a reflexive relation, since any geometric figure is congruent to itself. On the other hand, "is the son of" is not reflexive, since no person can be his own son, and the relation "understands" may or may not be reflexive, depending on whether a given individual understands himself.

A relation R is said to be *symmetric* if, for any two elements x and y of a given set, the relation xRy always implies the relation yRx. For example, the relation "is married to" is symmetric since, if x is married to y, then y is married to x. So, too, are the geometric relations of perpendicularity and parallelism. But the relation "is greater than" as applied to numbers is not a symmetric relation, nor is the relation "is a brother of" as applied to people since x may be the brother of y, but y may be the sister of x.

A *transitive* relation is one in which, if

$$xRy \quad \text{and} \quad yRz,$$
then
$$xRz$$

Thus, the relation "is a subset of" is transitive since, if x is a subset of y and y is a subset of z, then x is a subset of z. Other transitive relations are "is greater than" and "is less than" as applied to numbers, "is congruent to" as applied to geometric figures, and "is parallel to" as applied to straight lines. But the relation "can defeat" as applied to teams in athletic contests is not transitive, since it frequently happens that x can defeat y, and y can defeat z, but x cannot defeat z. Nor are the following relations transitive: "is perpendicular to" as applied to straight lines, "is the square of" as applied to numbers, and "is the daughter of" as applied to people.

Axioms for Equality

Returning now to the consideration of equality, we can state with some meaning that the relation of equality satisfies the following axioms for all elements a, b, and c of a given set.

Axiom 1. Equality is reflexive, that is, $a = a$.
Axiom 2. Equality is symmetric, that is, if $a = b$, then $b = a$.
Axiom 3. Equality is transitive, that is, if $a = b$ and $b = c$, then $a = c$.

Two additional axioms of equality are useful. They are concerned with the addition and multiplication of equalities.

Axiom 4. If $a = b$, then $a + c = b + c$.
Axiom 5. If $a = b$, then $a \cdot c = b \cdot c$.

It will be recognized that the last two axioms merely express in algebraic language the familiar statement "If equals are added to, or multiplied by, the same quantity, the results are equal."

These five axioms are the basis of many simple algebraic manipulations. Although such manipulations are usually made routinely, it is important to recognize their dependence on the system of axioms. It is only through such recognition that a true understanding of the basic operations of algebra is possible. The following examples, together with Problems 13–22 of Exercise 16, illustrate how these axioms may be used to support some of the elementary but fundamental algebraic processes.

106   What is the Axiomatic Basis of Algebra?

EXAMPLE 1. If $s = kt$, then $kt = s$ (axiom of symmetry).

EXAMPLE 2. If $2y = 7$, then $2y \cdot 1/2 = 7 \cdot 1/2$ (axiom of multiplication).

EXAMPLE 3. If $x = y$ and $y = 4$, then $x = 4$ (axiom of transitivity).

EXAMPLE 4. Show that if equals are multiplied by equals, the results are equal, that is, if $a = b$ and $c = d$, then $a \cdot c = b \cdot d$.
Solution:

$$
\begin{aligned}
a &= b & \text{(given)} \\
a \cdot c &= b \cdot c & \text{(axiom of multiplication)} \\
c &= d & \text{(given)} \\
b \cdot c &= b \cdot d & \text{(axiom of multiplication)} \\
a \cdot c &= b \cdot d & \text{(axiom of transitivity)}
\end{aligned}
$$

EXERCISE 16

THE EQUALITY RELATION

For each of the following relations, indicate whether the relation is (a) reflexive, (b) symmetric, or (c) transitive (Problems 1–10).

1. The relation "is the father of" for persons.
2. The relation ">" for real numbers.
3. The relation "is the square of" for real numbers.
4. The relation "is perpendicular to" for straight lines in a plane.
5. The relation "is similar to" for geometric figures.
6. The relation "belongs to the same political party as" for persons.
7. The relation "is a multiple of" for positive integers.
8. The relation "is a friend of" for people.
9. The relation "is in the same constellation as" for stars in space.
10. The relation "is supplementary to" for angles formed by straight lines.
11. A relation that is reflexive, symmetric, and transitive is called an *equivalence relation*. Which of the relations listed in Problems 1–10 is an equivalence relation?
12. Is congruence an equivalence relation? Explain your answer.

Indicate the axiom of equality that justifies each of the following statements.

13. If $E = IR$, then $IR = E$
14. $y = y$
15. If $y = 2x$ and $2x = 3$, then $y = 3$
16. If $x = y$, then $2x = 2y$
17. If $2a - 3 = 15$, then $2a = 18$
18. $5x - 1 = 5x - 1$
19. If $v = 1/2gt^2$, then $2v = gt^2$
20. If $C = 2\pi r$, then $2\pi r = C$
21. If $5 = 2\pi r^2 + 2\pi rh$, then $5 - 2\pi r^2 = 2\pi rh$

22. If $D = 1.320\sqrt{h}$ and $h = 26$, then $D = 1.320\sqrt{26}$

Give one nonmathematical example of a relation R such that:

23. If xRy, then yRx.   **24.** If xRy and yRz, then xRz.
25. xRx
26. If xRy, then $y\not R x$ ($\not R$ means "not related to")
27. If xRy and yRz, then $x\not R z$ **28.** $x\not R x$
29. Using the five listed axioms for equalities, show that if $a = b$ and $c = d$, then $a + b = c + d$.

17. FIELD AXIOMS

In addition to the axioms for equality, algebra is based on certain other axioms which describe the properties of operations on the elements of the system. In ordinary algebra these elements are numbers, the operations are addition and multiplication, and the properties are among those previously discussed in our consideration of various number systems. More specifically, the number system of algebra exhibits the following properties for the two given operations:

    Closure
    Commutativity
    Associativity
    Distribution of addition with respect to multiplication
    Existence of identity elements
    Existence of inverse elements

A mathematical system for which these properties hold true is called a *field*. Therefore, the number system of algebra is a field. For purposes of reference we list the formal definition of a field as follows.

A field is any mathematical structure consisting of a set of elements S, along with an operation of addition denoted by $+$ and an operation of multiplication denoted by $\times$, satisfying the following axioms.

    1. If a and b are in S, so is $a + b$ (closure axiom for addition).
    2. If a and b are in S, so is $a \times b$ (closure axiom for multiplication).
    3. If a and b are in S, then $a + b = b + a$ (commutative axiom for addition).
    4. If a and b are in S, then $a \times b = b \times a$ (commutative axiom for multiplication).
    5. If a, b, and c are in S, then $a + (b + c) = (a + b) + c$ (associative axiom for addition).

108   *What is the Axiomatic Basis of Algebra?*

6. If a, b, and c are in S, then $a \times (b \times c) = (a \times b) \times c$ (associative axiom for multiplication).

7. There is an element in S called 0, such that $0 + a = a$ for all a in S (existence of identity element for addition).

8. There is an element in S called 1, such that $1 \times a = a$ for all a in S (existence of identity element for multiplication).

9. For each a in S there exists an element x such that $a + x = 0$ (existence of inverse elements for addition).

10. For each a in S, $a \neq 0$, there exists an element y in S such that $a \times y = 1$ (existence of inverse elements for multiplication).

11. If a, b, and c are in S, then $a \times (b + c) = (a \times b) + (a \times c)$ (distributive axiom).

We have already encountered several examples of number fields in this book. For example, each of the following constitutes a field.

(a) The set of real numbers, with ordinary addition and multiplication.

(b) The set of rational numbers, with ordinary addition and multiplication.

(c) The set of clock numbers $\{0, 1, 2, 3, 4\}$, with addition defined as a forward movement of the indicator and multiplication defined as repeated addition.

On the other hand, we have also considered several number structures which fail to qualify as fields. For example:

(d) The set of natural numbers. Note that this set fails to satisfy the axioms requiring the existence of an additive identity element and the existence of inverse elements for addition and multiplication. All other axioms are satisfied.

(e) The set of integers. This set satisfies all the axioms for a field except the existence of inverse elements for multiplication.

EXAMPLE 5. Show that a set of clock numbers $\{0, 1, 2, 3\}$, with addition defined as a forward movement of the indicator, and multiplication defined as repeated addition, does not form a field.

Solution: The following tables list all possible arrangements of binary operations, and the result of each operation:

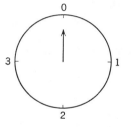

+	0	1	2	3
0	0	1	2	3
1	1	2	3	0
2	2	3	0	1
3	3	0	1	2

×	0	1	2	3
0	0	0	0	0
1	0	1	2	3
2	0	2	0	2
3	0	3	2	1

We note that:

(a) The closure axioms are satisfied. (Each result in the tables is an element of the set.)

(b) The commutative axioms are satisfied. (The listings in the tables are symmetric with respect to the diagonal from upper left to lower right.)

(c) The associative and distributive axioms are satisfied. (Verifiable by actual trials.)

(d) There exist identity elements for both operations. (One row and one column in each table duplicate the given elements.)

(e) There exists an additive inverse for each element in the set. (There is a zero result in the body of the table in each row and column.)

(f) There *does not* exist a multiplicative inverse for each element a in the set, $a \neq 0$. (Every row and column, except that composed of all zeros, would have to contain the element 1. For example, there is no element x such that $2 \cdot x = 1$.)

Therefore, the given set is not a field because the axiom of multiplicative inverses is not satisfied.

It is of passing interest to observe that a system of clock numbers forms a field when there are five elements in the set but does not form a field when there are four elements in the set. This is a particular example of a principle that the clock numbers form a field whenever the number of elements in the set is prime.

EXERCISE 17

Field Axioms

List the additive inverse and the multiplicative inverse (if they exist in the given set) for each of the following elements.

1. The element 7 in the set of natural numbers.
2. The element -2 in the set of integers.
3. The element $2/3$ in the set of rational numbers.
4. The element $\sqrt{3}$ in the set of real numbers.

Given the set $S = \{0, 1, 2\}$ with addition and multiplication defined as follows:

+	0	1	2
0	0	1	2
1	1	2	0
2	2	0	1

×	0	1	2
0	0	0	0
1	0	1	2
2	0	2	1

In Problems 5–15, if the given field axiom is true for all elements in the set S, give two specific examples that illustrate the axiom. If it is not true for all elements in the set S, give one example that illustrates that the axiom does not hold true.

5. Closure axiom for addition.
6. Closure axiom for multiplication.
7. Commutative axiom for addition.
8. Commutative axiom for multiplication.
9. Associative axiom for addition.
10. Associative axiom for multiplication.
11. Distributive axiom.
12. Existence of an identity element for addition.
13. Existence of an identity element for multiplication.
14. Existence of an additive inverse for each element in S.
15. Existence of a multiplicative inverse for each element in S except 0.
16. Does the system defined for set S in Problems 5–15 represent a field? Explain your answer.
17. Given the set of clock numbers $K = \{0, 1, 2, 3, 4, 5\}$, with addition defined as a forward movement of the indicator and multiplication defined as repeated addition. Complete an addition and multiplication table for the system, and use it to determine whether the field axioms hold true. Is this system a field? Explain your answer.
18. Does zero have (a) an additive inverse? (b) a multiplicative inverse? Explain.
19. If x is any positive integer, is the set $S = \{x + \sqrt{2}\}$ a field under the operations of addition and multiplication? Explain your answer.
20. If x and y are any two integers, is the set $S = \{x + y\sqrt{3}\}$ a field under the operations of addition and multiplication? Explain your answer.

18. ALGEBRAIC EXPRESSIONS AND OPERATIONS

The field axioms are the basis for all the fundamental operations involving algebraic expressions: collecting terms, factoring, simplifying expressions, and so on. We will now illustrate that this is so. The material considered at this point is of a review nature and, therefore, no more time than is necessary to recall manipulative skills will be devoted to it. Our concern will be with the justification of algebraic processes rather than with the memorization of techniques.

Addition and Subtraction of Algebraic Expressions

The addition of two similar terms, such as $7y$ and $5y$, is based on the distributive axiom, as follows.

EXAMPLE 1. $7y + 5y = (7 + 5)y = 12y$. This procedure is readily extended to the operation of subtraction or to the combining of more than two terms, as illustrated in the following examples:

EXAMPLE 2. $4x - 7x = [4 + (-7)]x = -3x$.

EXAMPLE 3. $2k + 3k + k = (2 + 3 + 1)k = 6k$.

It will be recalled that single algebraic terms such as the addends in these examples are known as *monomials*, while algebraic expressions that consist of a single term or the indicated sum of two or more single terms are known as *polynomials*. Polynomials of two or three terms occur with some frequency; they are called *binomials* and *trinomials*, respectively.

The procedure for adding and subtracting polynomials is supported by the application of the commutative, associative, and distributive axioms. Consider, for example, the addition of the binomials $3a + 5b$ and $4a - 2b$.

EXAMPLE 4. $(3a + 5b) + (4a - 2b)$
$= (3a + 5b) + (-2b + 4a)$ (commutative property of addition)
$= 3a + (5b - 2b) + 4a$ (associative property of addition)
$= 3a + 3b + 4a$ (distributive axiom)
$= 3a + 4a + 3b$ (commutative property of addition)
$= (3a + 4a) + 3b$ (associative property of addition)
$= 7a + 3b$ (distributive axiom)

Once the basis for the procedure is understood, of course, it is not necessary to perform all the steps. The following examples illustrate commonly accepted methods which result from these considerations.

EXAMPLE 5. Combine the expressions $2x - 7y - z$ and $6x + 3y - 5z$.
$(2x - 7y - z) + (6x + 3y - 5z) = (2x + 6x) + (-7y + 3y) + (-z - 5z)$
$= 8x - 4y - 6z$

EXAMPLE 6. Subtract $(3x^2 - 7x + 2)$ from $8x^2 - 9$.
$(8x^2 - 9) - (3x^2 - 7x + 2) = 8x^2 - 9 - 3x^2 + 7x - 2$
$= 5x^2 + 7x - 11$

Exponential Notation

Before proceeding with the remaining algebraic operations, it is necessary to recall the symbolism associated with repeated multiplication. Just as $b + b + b + b$ can be written as $4b$, the expression $b \cdot b \cdot b \cdot b$ can be written as b^4. In the expression b^4, b is called the *base* and 4 is called the *exponent.* The general expression b^n is read "the *n*th power of *b*" and is defined as follows:

$b^n = b \cdot b \cdot b \cdots b$ to *n* factors, where *n* is a natural number

If $n = 3$, we also refer to b^n as the *cube of b;* if $n = 2$, then b^n is also called the *square of b;* if $n = 1$, we usually write b^n merely as *b*.

Among the consequences of this definition are the following important relations, known as *laws of exponents,* where *a* and *b* denote real numbers ($b \neq 0$) and *m* and *n* denote positive integers.

1. $b^m \cdot b^n = b^{m+n}$, for example, $x^2 \cdot x^5 = x^7$
2. $(b^m)^n = b^{mn}$, for example, $(y^3)^4 = y^{12}$
3. $(ab)^m = a^m b^m$, for example, $(2k)^3 = 8k^3$
4. $b^m/b^n = b^{m-n}$ if $m > n$, for example, $t^8/t^2 = t^6$
   $= 1/b^{n-m}$ if $m < n$, for example, $v^5/v^9 = 1/v^4$
   $= 1$ if $m = n$, for example, $w^4/w^4 = 1$

These laws are a consequence of the properties of real numbers and the given definition of an exponent. For example, the law $b^m \cdot b^n = b^{m+n}$ may be derived by writing *m* factors of *b*, followed by *n* factors of *b*. The product is obviously $m + n$ factors of *b*. Similar arguments can be used to derive the remaining laws.

Multiplication of Algebraic Expressions

The multiplication of monomials, such as $-3x \cdot 2y$, is effected by applying the commutative property of multiplication to rearrange the factors and then applying the associative property of multiplication to simplify the expression. Thus

$$(-3x)(2y) = -3 \cdot 2 \cdot x \cdot y = -6xy$$

This simple procedure, together with the distributive law, permits us to multiply any polynomial by a monomial. For example,

$$4a(a^2 - 2a + 5) = 4a^3 - 8a^2 + 20a$$

Although it is not easily apparent, the same principles are the basis for multiplication when both factors are polynomials. Thus, to find the product of two binomials, the distributive property is used twice, as follows.

EXAMPLE 7.

$$\begin{aligned}(2x - 3a)(5x + 4a) &= (2x - 3a)5x + (2x - 3a)4a \quad \text{(distributive property)} \\ &= 5x(2x - 3a) + 4a(2x - 3a) \quad \text{(commutative property)} \\ &= 10x^2 - 15ax + 8ax - 12a^2 \quad \text{(distributive property)} \\ &= 10x^2 - 7ax - 12a^2\end{aligned}$$

It is evident that the product of two polynomials is the sum of all partial products obtained by multiplying each term of one polynomial by each term of the other. Therefore, the work can be conveniently arranged as follows:

$$
\begin{array}{r}
2x - 3a \\
5x + 4a \\
\hline
10x^2 - 15ax \\
8ax - 12a^2 \\
\hline
10x^2 - 7ax - 12a^2
\end{array}
$$

A final refinement is to write the product of two binomials in horizontal form by performing the multiplications in the order indicated by the arrows:

$$(2x - 3a)(5x + 4a) = 10x^2 - 7ax - 12a^2$$

The first term of the product is obtained by multiplying the first terms of the binomials, and the last term of the product is found by multiplying the last terms of the binomials. The middle term of the product is the sum of the partial products formed by multiplying the first term of each binomial by the last term of the other binomial. Applications of this procedure are many and varied. The following are some typical examples.

EXAMPLE 8. $(3x - 2)^2 = (3x - 2)(3x - 2) = 9x^2 - 12x + 4$.

EXAMPLE 9. $(3\sqrt{2} + 7)(2\sqrt{2} + 3) = 12 + 23\sqrt{2} + 21 = 33 + 23\sqrt{2}$.

EXAMPLE 10. $(7v^2 + 2)(v - 3) = 7v^3 - 21v^2 + 2v - 6$.

EXAMPLE 11. $(5y - 3)(5y + 3) = 25y^2 - 9$.

Division of Algebraic Expressions

Since division and multiplication are inverse operations, the division of an algebraic expression by a monomial divisor follows directly from a consideration of the corresponding multiplication. For example,

$$12x^5 \div 4x^2 = 3x^3 \quad \text{and} \quad 12x^5 \div 3x^3 = 4x^2$$

because

$$4x^2 \cdot 3x^3 = 12x^5$$

Similarly,

$$(6y^3 - 14y^2 + 2y) \div 2y = 3y^2 - 7y + 1$$

because

$$2y(3y^2 - 7y + 1) = 6y^3 - 14y^2 + 2y$$

What is the Axiomatic Basis of Algebra?

The division of one polynomial by another is analogous to the "long division" algorithm for integers. The algorithm for the division of A by B is based on the following principle.

> For any two expressions A and B with $B \neq 0$, there exist two expressions Q and R such that $A = B \cdot Q + R$.

Q is called the quotient, and R is called the remainder. If $R = 0$, then Q and B are said to be exact divisors of A.

The procedure for determining the quotient Q and remainder R for a specific case is illustrated in the following example.

EXAMPLE 12.

$$
\begin{array}{r}
3x^2 - 4x + 2 \\
2x - 5 \overline{\smash{\big)}\, 6x^3 - 23x^2 + 24x - 3} \\
\underline{6x^3 - 15x^2} \\
-8x^2 + 24x \\
\underline{-8x^2 + 20x} \\
4x - 3 \\
\underline{4x - 10} \\
7
\end{array}
$$

with annotations: $6x^3 \div 2x = 3x^2$; $-8x^2 \div 2x = -4x$; $4x \div 2x = 2$; $3x^2(2x - 5)$; $-4x(2x - 5)$; $2(2x - 5)$; $7 \leftarrow$ Remainder.

Therefore, the quotient is $3x^2 - 4x + 2$ and the remainder is 7. This result may be verified by showing that

$$6x^3 - 23x^2 + 24x - 3 = (2x - 5)(3x^2 - 4x + 2) + 7$$

Factoring

The purpose of factoring an algebraic expression is to undo the process of multiplication. For example, since $(x - 3)(2x + 5) = 2x^2 - x - 15$, then

$$2x^2 - x - 15 = (x - 3)(2x + 5)$$

Expressed in this way, the expression $2x^2 - x - 15$ is said to be factored, and the expressions $x - 3$ and $2x + 5$ are called its factors. This is not a trivial manipulation—factoring is an important element in the simplification of algebraic expressions and in the solution of quadratic and higher degree equations.

It is obvious that the procedures for factoring are not as automatic as those for finding a product, and a little reflection will show that they are strongly dependent on a knowledge of how the terms of a product are developed. Consider, for example, the integral factors of the base five numeral 22. By repeated trials we can discover that $22_5 = 4_5 \cdot 3_5$, but we can write the factors directly only if we know and remember that $4_5 \cdot 3_5 = 22_5$. Thus, the ability to factor is closely related to the ability to form products. Although the process of factoring a general polynomial expression can be quite complicated, certain types of elementary products lend themselves to a systematic approach. We will review three such cases.

1. *Expressions having a common factor.* When all terms of a polynomial have a factor in common, the expression is factored by using the distributive law in reverse. The result will be two factors: the common factor and the expression obtained by dividing each term of the given polynomial by the common factor.

EXAMPLE 13. $2a - 6 = 2(a - 3)$.

EXAMPLE 14. $3x^2 - 6x + 15 = 3(x^2 - 2x + 5)$.

EXAMPLE 15. $v^3 - 3v^2 + 7v = v(v^2 - 3v + 7)$.

2. *Binomials of the form $a^n \pm b^n$.* The most common of these are the following standard forms. In each case the factorization is the result of experience with the corresponding multiplications, and each can be verified by performing the indicated multiplications:

$$a^2 - b^2 = (a - b)(a + b)$$
$$a^3 - b^3 = (a - b)(a^2 + ab + b^2)$$
$$a^3 + b^3 = (a + b)(a^2 - ab + b^2)$$

Note that no attempt is made to factor the expression $a^2 + b^2$. Although factorization is possible, it cannot be done within the set of real numbers and, therefore, is not considered at this point. The use of these standard forms in factoring certain binomials is illustrated in the following examples.

EXAMPLE 16. $4x^2 - 9 = (2x - 3)(2x + 3)$.

EXAMPLE 17. $y^3 - 8z^3 = (y - 2z)(y^2 + 2yz + 4z^2)$.

EXAMPLE 18. $h^6 + k^3 = (h^2 + k)(h^4 - h^2k + k^2)$.

3. *Trinomials of the form $ax^2 + bx + c$.* We know that if we multiply $(3x + 7)(2x - 1)$, the result, $6x^2 + 11x - 7$, is a trinomial of the given

form. The problem is to determine how to reverse the procedure if we do not know the original factors or, indeed, that such factors even exist. We begin with the assumption, based on our experience with multiplication, that the factors are binomials. Therefore we write

$$6x^2 + 11x - 7 = (\quad\quad)(\quad\quad)$$

Since the product of the first terms of the binomials is $6x^2$, the only possibilities are $3x \cdot 2x$ or $6x \cdot x$. The product of the second terms is -7, so the possibilities are $-7 \cdot 1$ or $7 \cdot (-1)$. We must now choose the first terms and last terms in such a way that they yield the correct middle term. (If none of them yields the given middle term, the expression is not factorable.) The complete list of possibilities is

$$(3x - 7)(2x + 1)$$
$$(3x + 7)(2x - 1)$$
$$(6x - 7)(x + 1)$$
$$(6x + 7)(x - 1)$$

Note that of these, only the factors $(3x + 7)(2x - 1)$ produce the given middle term. In practice, not all possibilities need to be tried. Often the magnitude of the middle term and the signs of the terms will narrow the list of possible factors.

Prime Factors

If a given polynomial P has no factors other than P, $-P$, $+1$, -1, it is said to be *prime*. In order to express an algebraic expression in terms of its prime factors, it is sometimes necessary to factor its initial factors again. For example,

$$12c^2 - 60c + 75 = 3(4c^2 - 20c + 25)$$
$$= 3(2c - 5)(2c - 5)$$
$$= 3(2c - 5)^2$$

Note that the same result is obtained (with a little more work) by first factoring the given expression into binomial factors:

$$12c^2 - 60c + 75 = (2c - 5)(6c - 15)$$
$$= (2c - 5)(3)(2c - 5)$$
$$= 3(2c - 5)^2$$

EXERCISE 18
ALGEBRAIC EXPRESSIONS AND OPERATIONS

Using exponents, write each of the following expressions in its most concise form.

1. $2aabbb$
2. $-uvvv$
3. $16xxxyy$
4. $7hkhkh$
5. $-4pqqr$
6. $-2aabbbc$

By writing out the repeated factors, show that:

7. $x^3 \cdot x^5 = x^8$
8. $m^6/m^2 = m^4$
9. $(r^2)^3 = r^6$
10. $(ab)^m = a^m b^m$
11. $(b^m)^3 = b^{3m}$
12. $b^3/b^5 = 1/b^2$

Find the sum of each of the following:

13. $2a - b + c$
    $5a + b + 2c$
14. $3x^2 + 2x - 7$
    $5x^2 - 7x + 8$
15. $u^3 - 3u^2v + 5uv$
    $4u^3 - 2u^2v - uv$

16. $(m^2 + 3mn - 2n^2) + (2m^2 - 3mn + 3n^2)$
17. $(6x^2 - 2y^2) + (x^2 + y^2) + (2x^2 + 9y^2)$
18. $(p + 2q + r) + (8q - 2r - 5p) + (6q - 8r)$

Subtract as indicated.

19. $(3a^2 + 4a - 1) - (a^2 + 5a + 2)$
20. $(4x^2 - 2xy + 3y^2) - (3x^2 - 2xy - y^2)$
21. $(pq + p^3 - 3q^2) - (p^3 + 6pq + 8q^2)$
22. From $y^2 - 7y + 2$ subtract $4y^2 + y + 1$.
23. Subtract $a^2 - 3$ from $3a^2 + 6a + 2$.
24. What must be added to $2c - 7$ to produce $6c - 3$?
25. What must be added to $5a$ to produce $-2b$?

Find the following products and check by substituting specific numbers for the variables.

26. $(x + 4)(x + 2)$
27. $(y - 7)(y - 3)$
28. $(u - 2)(u + 5)$
29. $(3v - 2)(v + 6)$
30. $(2a - 3b)(2a + 3b)$
31. $(7b + 1)(7b - 1)$
32. $(4k - 5)(4k - 5)$
33. $(3x - 2)(3x - 2)$
34. $(5x - 3y)(2x + 5y)$
35. $(8a + 3b)(3a - 7b)$

Using division, determine whether:

36. $(x - 2)$ or $(x - 3)$ are exact factors of $x^3 - x^2 - 14x + 24$.
37. $(y + 1)$ or $(y - 1)$ are exact factors of $y^3 + 2y^2 - y + 2$.
38. $(2v - 1)$ or $(v - 2)$ are exact factors of $4v^3 + 4v^2 - 7v + 2$.
39. $(3a + 1)$ or $(3a + 2)$ are exact factors of $9a^3 + 9a^2 - 4a - 4$.

Perform the indicated operations and simplify the result.

40. $(x + h)^2 + (x - h)^2$
41. $(2x + h)^2 + (2x - h)^2$
42. $(x + h)^2 + (x + h) - (x^2 + x)$
43. $(x + h)^2 + 2(x + h) - (x^2 + 2x)$
44. $\dfrac{3(x + h)^2 + 2(x + h) - (3x^2 + 2x)}{h}$
45. $\dfrac{2(x + h)^2 + 5(x + h) + 6 - (2x^2 + 5x + 6)}{h}$

Factor each of the following expressions by removing the common monomial factor.

46. $3x - 12$
47. $a^3 - a^2$
48. $\pi R^2 - \pi r^2$
49. $\pi r^2 + 2\pi r$
50. $ax + ay - a$
51. $4c^2 + 8c - 2$
52. $6x^2 + 9x - 12$
53. $2\pi R - 2\pi r$
54. $2\pi R h - 2\pi r h$

Factor each of the following binomials.

55. $a^2 - 9$
56. $y^2 - 25$
57. $9u^2 - 4v^2$
58. $16x^2 - 49y^2$
59. $a^3 - 1$
60. $x^3 + 8$
61. $8v^3 + y^3$
62. $h^3 - 27$
63. $k^6 - 8$

Factor each of the following trinomials.

64. $a^2 + 4a + 3$
65. $x^2 - 3x + 2$
66. $3y^2 + 8y + 4$
67. $2k^2 + 7k + 3$
68. $a^2 + 22a + 121$
69. $16x^2 + 8x + 1$
70. $6v^2 + 13v - 5$
71. $7b^2 - 41b - 6$
72. $2a^2 - 5ab - 3b^2$
73. $8m^2 + 10mn - 3n^2$
74. $p^2 - 20p + 100$
75. $6x^2 + 71x - 12$

Express each of the following formulas in factored form.

76. $A = \frac{1}{2}ah + \frac{1}{2}bh$ — Area of a trapezoid
77. $S = \frac{1}{2}a + \frac{1}{2}b + \frac{1}{2}c$ — Semiperimeter of a triangle
78. $x^2 - 2hx + h^2 = 4ay - 4ak$ — Equation of a parabola
79. $a^2 = c^2 - b^2$ — Pythagorean theorem
80. $S = 2\pi r^2 + 2\pi r h$ — Surface of a right circular cylinder
81. $S = 2ab + 2bc + 2ac$ — Surface of a rectangular solid
82. $A = P + Prt$ — Simple interest
83. $x^2 - 2hx + h^2 + y^2 - 2ky + k^2 = R^2$ — Equation of a circle
84. $S = n^2 + n$ — Sum of the first n even integers

Write each of the following expressions as a product of prime factors.

85. $a^3 - a^2 - 30a$
86. $2x^2 - 6x - 140$
87. $2x^6 + 2$
88. $x^4 - x^8$
89. $p^4 - 4p^2 + 4$
90. $4y^4 - 13y^2 + 9$

19. EQUATIONS AND INEQUALITIES IN ONE VARIABLE

We have made frequent use of letter symbols such as x or y to represent any element of a given set. Used in this way the symbol is called a *variable* and the given set is called the *domain* of the variable. As a limiting but important case, when the domain consists of a single element, the symbol is called a *constant*.

Consider now the algebraic expression $ax + b$, where a and b are constants and x is a variable whose domain is the set of real numbers. Examples of this expression are

$$7x + 2, \quad \tfrac{1}{2}x, \quad -4, \quad x - 7.3, \quad 0, \quad -5x + \tfrac{1}{4}$$

Each of these is called a *linear form* in the variable x. (Note that either or both of the constants a and b may be zero.) If the variable x is replaced by a number in its domain, the resulting number is called a *value* of the expression. For example, if x is replaced by 9 in the linear form $3x - 5$, the value of the expression is 22.

Two linear forms may be connected to form a relation. The relation may be expressed through the use of symbols as:

$=$ (is equal to)
$\neq$ (is not equal to)
$>$ (is greater than)
$<$ (is less than)
$\geq$ (is greater than or equal to)
$\not<$ (is not less than)

and so on. Our concern with such relations is to determine replacements of the variable that result in a true statement. Thus, for the relation

$$2x + 3 > x + 8$$

a replacement value of x which results in a true statement is $x = 6$. Another is $x = 9$. Each such replacement is called a *solution* of the relation, and the set of all replacements is called the *solution set* for the relation. To *solve* the relation means to determine its solution set.

Linear Equations in One Variable

A relation that states the equality between two linear forms, such as $3x - 1 = 5x + 4$, is called a *first degree equation* or a *linear equation* in that variable. The expression $3x - 1$ is called the left member of the equation, and $5x + 4$ is called the right member. To determine the solution set of a linear equation, the usual procedure is to apply the axioms of equality (Section 16) to transform the given relation to progressively simpler equations until the solution set is explicitly stated. This is illustrated in the following example.

EXAMPLE 1. Determine the solution set for the relation $3x - 1 = 5x + 4$.
Solution:

$3x - 1 = 5x + 4$

$-2x - 1 = 4$ Adding $-5x$ to each member

$-2x = 5$ Adding 1 to each member

$x = -5/2$ Dividing each member by -2

Linear Inequalities in One Variable

The solution set for an inequality of two linear forms is determined by a method analogous to that used for the solution of a linear equation. It is based on the following principles.

1. If the same number is added to (or subtracted from) both members of an inequality, the results are unequal in the same order. The expression "same order" means that both inequalities are "greater than" or that both are "less than." The expression "same sense" is frequently used instead of "same order." The following examples illustrate this principle:

(a) $7 > 3$ (b) $6 > 4$ (c) $-2 < 1$

$7 + 5 > 3 + 5$ $6 + (-3) > 4 + (-3)$ $-2 + 8 < 1 + 8$

$12 > 8$ $3 > 1$ $6 < 9$

2. If both members of an inequality are multiplied (or divided) by the same *positive* number, the results are unequal in the same order. For example:

(a) $8 > 7$ (b) $-6 < -2$

$8 \cdot 3 > 7 \cdot 3$ $-6 \div 2 < -2 \div 2$

$24 > 21$ $-3 < -1$

3. If both members of an inequality are multiplied (or divided) by the same *negative* number, the results are unequal in the opposite order.

For example:

(a) $\quad 3 > 2$
$\quad 3(-2) < 2(-2)$
$\quad -6 < -4$

(b) $\quad -2 > -9$
$\quad -2(-1) < -9(-1)$
$\quad 2 < 9$

(c) $\quad -9 < -6$
$\quad -9 \div (-3) > -6 \div (-3)$
$\quad 3 > 2$

We now apply these principles to the solution of some simple inequalities. Note that when the domain of the variable is the set of real numbers, the solution set will usually be an infinite subset. This is in contrast to the solution set of a linear equation, which consists of either the entire set, the empty set, or a unit set.

EXAMPLE 2. Determine the solution set for the relation $6x - 2 > x + 13$.
Solution:
$\quad 6x - 2 > x + 13$
$\quad 5x - 2 > 13 \qquad$ Adding $-x$ to each member
$\quad 5x > 15 \qquad$ Adding $+2$ to each member
$\quad x > 3 \qquad$ Dividing each member by 5

EXAMPLE 3. Solve the inequality $2y + 3 > 5y - 9$.
Solution:
$\quad 2y + 3 > 5y - 9$
$\quad -3y + 3 > -9 \qquad$ Adding $-5y$ to each member
$\quad -3y > -12 \qquad$ Adding -3 to each member
$\quad y < 4 \qquad$ Dividing each member by -3

EXAMPLE 4. Solve the relation $8z + 3 \leq 9z - 5$.
Solution:
$\quad 8z + 3 \leq 9z - 5$
$\quad -z + 3 \leq -5 \qquad$ Adding $-9z$ to each member
$\quad -z \leq -8 \qquad$ Adding -3 to each member
$\quad z \geq 8 \qquad$ Dividing each member by -1

We may verify the solution by observing that for $z = 8$ the two members are equal, and for $z > 8$ the left member is less than the right member. Therefore, the relation is satisfied for $z \geq 8$. However, for $z < 8$ the left member is greater than the right member. Therefore, the relation is satisfied if, and only if, $z \geq 8$.

EXERCISE 19

EQUATIONS AND INEQUALITIES IN ONE VARIABLE

Describe the domain of a variable x, if x is:

1. The product of two integers.
2. Three less than a positive rational number.
3. Twice a positive integer.
4. The temperature on a centigrade scale.
5. The length of the side of a square.
6. The sum of two rational numbers.
7. The product of two irrational numbers.

If the domain of the given variable is the set of real numbers, determine the solution set for each of the following relations.

8. $3x + 1 = 7$
9. $2y - 3 = 4y + 1$
10. $x + 7 > 9$
11. $z - 3 < 0$
12. $5v - 2 = 8 + v$
13. $8 - 4t = 5t - 3$
14. $9x + 2 > 2 - 2x$
15. $2 - 3w > 4w + 5$
16. $7y + 7 < 5y - 9$
17. $5 - 8z < 2z + 15$
18. $u + 3 \geq 5u$
19. $2x \leq 6x - 12$
20. $r - 3(2 + r) = 6r + 1$
21. $2(v + 1) > 3(v + 2) - 5v$

If the domain of the variable is the set of positive rationals, determine the solution set for each of the following relations.

22. $4 + x = 2x - 8$
23. $3v + 2 < 5v - 6$
24. $4(t + 1) \geq 3t - 9$
25. $7y + 13 = 3 + 2y$
26. $8w + 1 \leq w - 13$
27. $2 - 3z < 2(2 - 3z)$
28. $3 + 3r > 4 + 4r$
29. $6y - 5 < 0$
30. $3(x + 7) = 4 - 2x$

List four distinct numbers in the solution set of the relation $x < 2$ if the domain of the variable is:

31. The set of integers.
32. The set of positive rationals.
33. The set of negative rationals.
34. The set of negative integers.
35. The set of positive irrationals.

Define the domain of the variables x and y such that:

36. $x > y$ and $x^2 > y^2$
37. $x > y$ and $x^2 < y^2$
38. $x > y$ and $1/x > 1/y$
39. $x > y$ and $1/x < 1/y$
40. $x > y$ and $x/2 > y/2$

If the domain of x is the set of real numbers, describe the relationship of the sets A and B (in terms of union, intersection, complements, etc.) if:

41. $A = \{x \mid x \leq 3\}$, $B = \{x \mid x > 3\}$
42. $A = \{x \mid x < 5\}$, $B = \{x \mid x < 7\}$
43. $A = \{x \mid x \geq 1\}$, $B = \{x \mid x \leq 0\}$
44. $A = \{x \mid x \leq 6\}$, $B = \{x \mid x \not> 6\}$

20. APPLICATIONS OF SIMPLE LINEAR RELATIONS

A critical step in dealing with an applied problem is to translate the conditions imposed by the problem into a symbolic mathematical statement. Once this is done, it becomes a fairly routine matter to analyze the resulting relation or to find its solution set. Since significant problems which are met in practice are likely to be complicated and technical, the problems that we consider will represent simplified situations. A consideration of such "verbal problems" or "word problems" is important, nevertheless, because it leads to the ability to handle important problems in specialized situations.

Applied problems which can be solved by the use of algebra are so varied that no single procedure can cover all situations. Experience and practice count heavily. It is possible, however, to follow certain general steps which will give direction to the analysis of a particular problem. The following examples illustrate these steps.

EXAMPLE 1. Three men agree to purchase an airplane, sharing the cost equally. By including an additional man in the venture, the cost to each will be reduced by $1250. What is the projected cost of the airplane?

Step 1. Read the problem for understanding. Depending on the situation, this may mean reading it several times, visualizing it, experimenting with specific numbers, drawing a diagram, or estimating an answer. In particular, it is necessary to identify the facts given in the problem and to determine what is required. The facts given in this problem are (a) initially, each man would pay one-third of the cost of the airplane and (b) this one-third share would be reduced by $1250 if a fourth man is added to the group. It is required to determine the cost of the plane.

Step 2. Represent each unknown quantity by a variable or an expression containing the variable. Be certain that the variable represents some measureable quantity. Describe it accurately. In the stated problem we can choose to let

x = cost of the airplane, in dollars (note that the domain of x is the set of positive real numbers)

then $\tfrac{1}{3}x$ = initial cost to each man if there are three partners

and $\tfrac{1}{4}x - 1250$ = cost to each man if there are four partners

Step 3. Write a mathematical relation based on the facts of the problem. In this problem we know that the cost to each of the four partners is $\tfrac{1}{4}x - 1250$ and the total cost is x. Therefore,

$$4(\tfrac{1}{4}x - 1250) = x$$

Step 4. Determine the solution set for the stated relation, and use the solution set to evaluate each unknown required by the problem. The relation developed in Step 3 is a simple linear equation whose solution set is determined as follows:

$$4(\tfrac{1}{3}x - 1250) = x$$
$$\tfrac{4}{3}x - 5000 = x$$
$$\tfrac{1}{3}x = 5000$$
$$x = 15{,}000$$

Therefore, the cost of the airplane is $15,000.

Step 5. Check the results by substituting in the original problem. If three men were to purchase the plane, each would pay $5000. If four men were to purchase the plane, each would pay $3750. Therefore, by including one additional man, the cost to each is reduced by $1250.

EXAMPLE 2. A professional baseball team has won 68 of its first 100 games. It is scheduled to play a 162 game season. How many of the remaining games must it win to finish the season with *at least* a 65% win record?

Step 1. Understanding this problem requires a knowledge that the win record is computed by dividing the number of games won by the number of games played and converting this figure to percent. The number of wins needed in the remaining 62 games to compile a season record of 65% or better must be determined.

Step 2. Let x = number of wins needed in the remaining 62 games. Note that the domain of x is the set of positive integers between 0 and 62.

Step 3. Since x is the number of wins required in the final 62 games and the team has already won 68 games, the total wins for the season will be $68 + x$. The total games will be 162. Therefore, the required relation is

$$\frac{68 + x}{162} \geq 0.65$$

Step 4. Solving the given relation, we have

$$\frac{68 + x}{162} \geq 0.65$$

$68 + x \geq 105.3$ Multiplying each member by 162

$x \geq 37.3$ Subtracting 68 from each member

$x \geq 38$ The domain of x is restricted to positive integers

Therefore, the team must win 38 or more of its remaining games in order to compile a season record of at least 65%.

Step 5. Check: If the team were to win 37 of its remaining games, its final record would be

$$\frac{68 + 37}{162} = \frac{105}{162} = 64.8\%, \quad \text{which is less than } 65\%$$

If the team were to win 38 of its remaining games, its final record would be

$$\frac{68 + 38}{162} = \frac{106}{162} = 65.5\%, \quad \text{which is more than } 65\%$$

It is clear that any number of wins greater than 38 will increase this result.

EXERCISE 20

Applications of Simple Linear Equations

Express each of the following in algebraic symbols.

1. x is less than 7.
2. y is not more than 7.
3. z exceeds 5 by n.
4. The product of u and v is w.
5. What number exceeds b by 3?
6. What number is 13 less than k?
7. v is c less than twice u.
8. w is greater than the sum of m and n.
9. h is not greater than q.
10. k is p greater than q.
11. What is the next consecutive integer after $2n - 1$?
12. If the sum of two numbers is 19 and one of the numbers is n, what is the other number?
13. Two numbers total N. If the smaller is y, what is the larger?

14. A cost of D dollars is shared equally by n individuals. What is the cost to each?
15. What is the cost of four tires at d dollars per tire?
16. How many hours are there in d days?
17. How many feet are there in i inches?
18. Don is x years old. How old was he nine years ago?
19. What distance does a car traveling at r miles per hour cover in $1\frac{1}{2}$ hours?
20. Two trains depart from the same station in opposite directions. If one travels at an average rate of r_1 miles per hour and the other at r_2 miles per hour, how far apart will they be after t hours?
21. A workman installs an average of n square feet of floor tile in h hours. How much tile can he install in three hours?
22. What is the total value, in cents, of d dollars and q quarters?
23. What is the total value, in cents, of q quarters, d dimes, and n nickels?
24. What is the simple interest on x dollars at 4% per year for 9 months?

Solve each of the following problems.

25. The two oldest employees, from the standpoint of service with a given firm, have an aggregate of 59 years of experience. If one has three years more seniority than the other, how many years has each been with the firm?
26. A 38 inch length of copper tubing is to be cut into two pieces. If one piece is to be 3 inches shorter than the other, how long should each piece be?
27. The president of a corporation earned $47,500 this year. This was at least $1000 more than twice the amount earned by any one of his district managers. What were the maximum earnings of a district manager?
28. Paul's father is three times as old as Paul. Twelve years from now he will still be more than twice as old as Paul will be. What do you know about Paul's present age?
29. The sum of three consecutive even integers is 138. Find the numbers.
30. Find two consecutive odd integers whose sum is 76.
31. For what solution set does the linear form $7x - 2$ have the same value as the linear form $2x + 5$?
32. For what solution set does the linear form $8 - 3x$ exceed the value of the linear form $4x - 6$?
33. If the side of a square is increased by 1 unit, its area is increased by 27 square units. Determine the size of the original square.
34. The length of a rectangular table top is 2 inches more than twice its width. If 154 inches of molding are needed to cover its outside edges, find the dimensions of the table top.
35. A deficit of $2,125,000 in the operation of a trade fair is to be prorated among three government agencies. If the county is to contribute 25% more than the city and the state is to contribute twice as much as the city, find the amount to be underwritten by each.

36. An investment of $15,000 in bonds, part at 5% and the balance at 6%, yields $860 annually. Find the amount invested at each rate.
37. A collection of $6.55 in quarters and dimes contains six more dimes than quarters. How many coins of each type are in the collection?
38. The daily payroll for a construction gang of 23 men is $815. If the skilled workers earn $45 per day and the unskilled workers earn $23 per day, find the number employed in each classification.
39. In an electric circuit the impressed voltage, E, is equal to the product of the current in amperes, I, and the resistance in ohms, R. What is the amount of current flowing through a 24 ohm resistor if the impressed voltage is 120 volts?
40. The velocity, v, of an object after t seconds is given by the relation $v = v_0 + at$, where v_0 is its initial velocity and a is the accelerative factor. What acceleration is being imparted to an object that starts at a velocity of 10 feet per second and is moving at 234 feet per second after 7 seconds?
41. How much water must be evaporated from 1000 gallons of a 12% salt solution if the residual solution is to contain 20% salt?
42. How much acid must be added to 100 cubic centimeters of an 8% solution to increase its concentration to at least 10%?

4

What are Some of the Fundamental Properties of Geometric Forms?

21. THE BEGINNINGS OF GEOMETRY

Elementary knowledge of geometric forms, such as the straight line and straight-line figures, circles, and solids such as the cube, dates back to antiquity. So, too, do practical applications of geometry such as measurement of land and simple measures of capacity such as the cubic foot. In fact, the development of geometry by the Egyptians is attributed to the necessity of reestablishing land boundaries washed away by periodic overflowing of the Nile. The word *geometry* is derived from the Greek language and literally means "land measure."

The great scholars of all ages were never content to view any body of scientific knowledge as a collection of facts. Scholars have always asked "how" and "why," and it is to Thales (640–550 B.C.) that we ascribe the origin of geometry as a deductive science rather than a useful collection of facts. It was Euclid, however, born in the third century B.C., who, in his *Elements*, presented geometry as a system of pure mathematics which still

Figure 4.1   **Figure 4.2**

remains a model of a deductive system. While modern mathematicians have, to be sure, found certain flaws in the logic of Euclid, the geometry of Euclid is essentially valid today.

In Section 2 of Chapter 1 we pointed out the fact that conclusions reached by deductive reasoning are frequently first discovered by inductive reasoning, that is, by observation. We have also seen that deductive proofs need to be supported by certain basic assumptions as well as by defined and undefined terms. Using only three simple (and undefined) elements of geometry, the point, the straight line, and the plane, we shall make observations leading to a few basic assumptions.

Figure 4.1 seems to indicate that through one point any number of straight lines may be drawn, and it also suggests that these lines may be drawn as long as we please. Figure 4.2 suggests that through two points only one straight line may be drawn, as long as we please. Figure 4.3 indicates that through three points chosen at random, it is, in general, not possible to draw one straight line. Figure 4.4, in which MN represents part of a plane, such as a table top, suggests that if any two points in a plane are joined by a straight line, the line will lie entirely in the plane. Now most students asked to verify the conclusions suggested by the first three figures would have little difficulty in doing so; the only "tools" needed would be a sheet of paper, a pencil, and a straight edge. A table top and a straightedge would serve nicely to verify the conclusion suggested by Figure 4.4; placing the straightedge in various positions on a table top would reveal that there were no humps or hollows under the rule. Every part of the straightedge would touch the table top, which represents the plane.

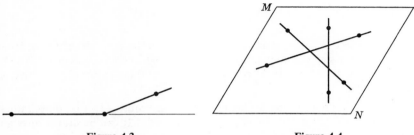

Figure 4.3   **Figure 4.4**

A more thoughtful student might ask, "What would I do if I did not have a straightedge, that is, a ready-made straight line?" Strangely enough, while a circle may be drawn quite accurately with a compass, which in no way *resembles* a circle, there is no simple way to draw a straight line without another straight line as a guide. The student, in seeking an answer to his question, might observe that the intersection of any two walls of his room certainly appears to be a straight line, as do the sides of a chalk box or any similar carton. To further pursue this idea, let us carefully fold a sheet of writing paper near the middle, using a thumb or a pencil to make the crease as sharp as possible. Then, using the crease as a straightedge to draw several lines, we would conclude that the crease is indeed a good representation of a straight line. It appears that by folding and creasing the paper, we have constructed a model of a straight line without the use of another straight line to guide us. But we have done much more than this; we have indicated a very plausible axiom. Assuming that each half of the folded paper is a plane, we may state as an axiom:

The intersection of two planes is a straight line.

In summary, we have introduced the following important properties of points, lines, and planes.

1. Any number of straight lines may be drawn through a single point.
2. A straight line can be produced continuously in either direction.
3. One and only one straight line can be drawn through two given points.
4. If any two points in a plane are joined by a straight line, the straight line lies entirely in the plane.
5. The intersection of two planes is a straight line.

EXERCISE 21

THE BEGINNINGS OF GEOMETRY

1. (a) On a sheet of paper, draw three points not on the same straight line. How many distinct straight lines can be drawn through them? (b) Repeat for 4, 5, 6, and 7 points. Be sure that no more than two of these points are on any one straight line.
2. When more than 4 or 5 points are considered in Problem 1, it becomes rather difficult to determine the number of straight lines that can be drawn through them by actually drawing these lines. Study and then complete the following table using inductive reasoning.

21. The Beginnings of Geometry

Number of Points	Number of Lines	Difference
3	3	
		3
4	6	
		4
5	10	
		5
6	15	
		6
7	21	
8		
9		
10		

3. Even with the use of the table in Problem 2, it would be tedious, if not difficult, to determine the number of lines through 50 or 100 points. Verify for $m = 6, 7, 8, 9$, and 10, that the formula

$$n = \frac{m(m-1)}{2}$$

is a correct formula for determining the number of straight lines, n, that can be drawn through m points, no more than two of which are on any one straight line. Find the number of straight lines when $m = 100$.

4. Supply the reason for each step of the following deductive proof of the formula in Problem 3.
   (1) From each of the m points, $m - 1$ lines can be drawn.
   (2) Therefore, the total number of lines drawn is $m(m - 1)$.
   (3) But only half of these lines are distinct.
   (4) Therefore, $m(m - 1)/2$ is the total number of distinct lines that can be drawn.

5. Complete the following table for the number of diagonals in a polygon.

Number of Sides	Number of Diagonals	Difference
3	0	
		2
4	2	
		3
5	5	
		4
6	9	
7		
8		
9		
10		

132   *What are Some of the Fundamental Properties of Geometric Forms?*

6. Try to discover a formula for the number of diagonals in a polygon of m sides.
7. Give a deductive proof of the formula you discovered in Problem 6. Follow the pattern of the proof in Problem 4.
8. The term "straightedge" has been used several times in this section. In what way does a straightedge differ from a ruler? What conclusion may be drawn from the fact that the term "ruler" was never used in Euclid's *Elements*?

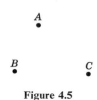

Figure 4.5

9. Let three points be given as in Figure 4.5. How many positions may a fourth point D assume, so that the figure $ABCD$ is a parallelogram? (Any other order of the four letters may be used.)
10. Sketch a surface such that a straight line joining some pair of its points, but not every pair, lies entirely in the surface.
11. Draw ten points in five straight lines so as to have four points in each line.
12. Draw 13 points in 12 straight lines so that there will be three points in each line.
13. Draw a triangle. Find the midpoint of each side, and draw a straight line from each midpoint to the opposite vertex. What theorem does this suggest? Have you used inductive or deductive reasoning in stating this theorem?
14. In Exercise 1, Problem 18, it was shown that the triangular numbers 1, 3, 6, 10, ... are all sums of consecutive integers. Draw a geometric figure to illustrate a possible reason for calling these numbers "triangular."
15. Draw a rhombus, that is, a parallelogram with all sides equal. Draw the diagonals. What property of the diagonals of a rhombus is suggested?

22. THE ELEMENTS OF GEOMETRY

We now consider in somewhat more detail three of the basic undefined terms of geometry: point, line, and plane. In doing so, it will be useful to contrast Euclid's treatment of these concepts with those presented in modern texts. It is not surprising that Euclid's work contained some logical imperfections, partly because it was a very early application of axiomatic methods. Therefore, it is no discredit to him that, some 2200 years later, more rigorous approaches to the foundations of geometry were developed.

The Point

Euclid attempted to define the word "point" as ". . . that which has no parts." The definition is obviously inadequate because the term "parts" is not a simpler term and the word "that" does not refer to any specified set of elements. Modern practice is to accept the term "point" as undefined. Our intuitive notions of what we mean by a point are evident in statements concerning the properties of points. We say that a point has no dimensions, that all geometric figures are considered to be sets of points, and that a point represents a position on a line, on a plane, or in space.

One consequence of these descriptions is that two points, A and B, may in certain contexts turn out to be one and the same point. If the intent is that two "different" points are to be considered, we may refer to them as "two *distinct* points A and B."

The Line

Euclid's definition of a line is, "A line is breadthless length." Again, we do not consider this description a valid definition because the concepts of "breadth" and "length" are not simpler terms. As in the case of the point, the concept of line is today accepted as an undefined term in geometry. Some of the important properties possessed by lines are the following.

1. A line is a set of points. The set can be considered generated by a point moving in a given path. In general, the path is called a *curve*, but if the point moves without changing direction the path is a straight line. Therefore, the word "curve" includes the straight line as a special case.

2. Two distinct points determine a line. This means that given two distinct points such as A and B in Figure 4.6, there exists one and only one straight line containing both A and B. We call this line AB and designate it as $\overleftrightarrow{AB}$, or by a single letter such as m. Implicit in this symbolism is the understanding that a line has no endpoints. The portion of the line between A and B, including A and B as endpoints, is called a *line segment*, and is commonly indicated by the symbol $\overline{AB}$.

3. On any line AB there exists at least one point between A and B. The word *between* is actually considered an undefined term, although its meaning may be intuitively clear. In Figure 4.6, the point C is between A and B. The point C (or any other point on the line) separates the given line into two

Figure 4.6

half-lines. In particular, point C determines three sets of points: the two half-lines and a unit set containing only the element C. The union of the point C and one of the half-lines formed by it is called a *ray*. Thus the ray CB, denoted by the symbol $\overrightarrow{CB}$, contains the endpoint C and the half-line through the point B. Similarly, $\overrightarrow{BC}$ is the ray containing the point B as an endpoint and the half-line through the point C. It is evident that the intersection of the rays CB and BC is the line segment CB, that is,

$$\overrightarrow{CB} \cap \overrightarrow{BC} = \overline{CB}$$

Angles

An *angle* is defined as the union of two rays having a common endpoint and not both on the same line. The restriction that both rays do not lie on the same line is introduced to avoid, for the present, considering a line as an angle. In Figure 4.7 the rays AC and AB form the illustrated angle which is called $\angle CAB$ (read, angle CAB) or $\angle BAC$, or simply $\angle A$ if no confusion results. The common endpoint A is called the *vertex* of the angle, and the two rays are called *sides* of the angle. In set notation the definition of an angle is embodied in the statement

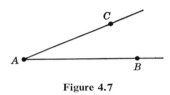

Figure 4.7

$$\overrightarrow{AC} \cup \overrightarrow{AB} = \angle CAB$$

Measurement of Line Segments and Angles

Length is a familiar but undefined property of line segments. Although Euclidean geometry is not concerned with the length of a line segment in terms of common units such as the inch or the yard, from a practical viewpoint the concept of length is not unimportant. A natural question that arises when we consider a line segment is, "How long is it?" We may define the measure of a line segment as the number of units contained in the segment. Recall that a segment with endpoints A and B is denoted by $\overline{AB}$; the measure of this segment is denoted by the symbol $m(\overline{AB})$. If the distance between points A and B is 6 inches, then $m(\overline{AB}) = 6$ when the length is measured in inches, and $m(\overline{AB}) = 1/2$ when the length is measured in feet.

The measure of an angle is given in terms of a unit angle just as the measure of a line segment is given in terms of a unit segment. Though several

Figure 4.8

common standards are used for a unit angle, the one used in everyday measurement is the degree. In Figure 4.8 consider the line AB and a point O on the line. If the line AB and the entire half-plane above it are divided into 180 equal parts by rays having the common endpoint O, then one of the angles formed is said to have a measure of 1 degree, written $1°$. Thus, if $\angle BOC$ is such an angle, it represents a unit angle. If the unit angle can be reproduced, let us say, 37 times in the interior of a given angle, then the measure of the given angle is 37 degrees.

The Plane

Suppose a physical representation of the angle illustrated in Figure 4.7 were made by soldering together the ends of two straight pieces of rigid wire. Would it lie on a flat surface, such as a desk top, in such a way that the two rays would touch the surface at all points? A little reflection should convince us that, in an ideal situation, it would do so. It is possible to visualize a flat surface passing through all points of the angle—such a surface is called a *plane*. In geometry, it is assumed that two distinct intersecting lines determine a plane.

Although planes are usually accepted as undefined, certain properties are commonly ascribed to them. Among these are:

1. A plane is a set of points. For any given point, the point is said to be *on* the plane and, conversely, the plane is said to pass *through* the point.

2. A plane is determined by any of these equivalent conditions: two distinct intersecting lines, three distinct points, or a line and a point not on the line.

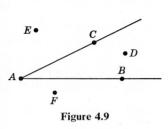

Figure 4.9

A plane is, therefore, completely identified by naming three points on it, although frequently two points or a single letter are sufficient to name a plane if no confusion results.

3. Any line on a plane separates the plane into two half-planes. This situation is analogous to the separation of a line by a point. The two half-planes and the line which separates the plane represent three disjoint sets of points. The concept of half-planes is useful in defining the concept of the interior of an angle.

In Figure 4.9 $\angle CAB$ determines a plane CAB, the line AB separates this plane into two half-planes, and the line AC separates the plane into another

136   *What are Some of the Fundamental Properties of Geometric Forms?*

pair of half-planes. If the half-plane determined by the line AB and containing C intersects the half-plane determined by the line AC and containing B, their intersection is called the *interior* of $\angle CAB$. Therefore, point D is said to be in the interior of $\angle CAB$, while points E and F are said to be in the exterior.

We have thus introduced various sets of points, such as lines, rays, angles, and planes. These sets can be considered as the basic building blocks of geometry, that is, other sets of points can be developed by using these sets together with the operations of intersection and union. One of our purposes in undertaking such developments is to gain an understanding of some newer geometries—the introduction of analytical geometry inherent in the study of functions in Chapters 5–8, and the discussion of finite geometry in Section 49.

EXERCISE 22

The Elements of Geometry

1. Given two distinct points M and N, illustrate and explain the distinction between the symbols $\overleftrightarrow{MN}$, $\overrightarrow{MN}$, and $\overline{MN}$.

In Problems 2–5, describe the set which is:

2. The union of two half-lines.
3. The intersection of two half-lines.
4. The union of two half-planes.
5. The intersection of two half-planes.
6. Given four points A, B, C, and D, which are not all in the same plane, how many planes do they determine?
7. Sketch three coplanar lines (lines on the same plane) which have (a) no point of intersection, (b) one point of intersection, (c) two points of intersection, and (d) three points of intersection.
8. Sketch three planes which have (a) no line of intersection, (b) one line of intersection, (c) two lines of intersection, and (d) three lines of intersection.

In Problems 9–15, state which of the following statements are true with reference to Figure 4.10.

Figure 4.10

9. $C \in \overline{AB}$ 10. $B \in \overline{AC}$ 11. $B \in \overrightarrow{AC}$
12. $\overline{AC} \cup \overline{AB} = \overline{AB}$ 13. $\overline{AC} \cup \overline{CB} = \overline{AB}$
14. $\overline{AC} \cap \overline{CB} = \phi$ 15. $\overline{AB} \cap \overline{CB} = \overline{CB}$
16. Triangles, quadrilaterals, pentagons, and hexagons are defined as polygons having 3, 4, 5, and 6 sides, respectively. A polygon is, in turn, defined as a plane figure having a specified number of sides. For which of these figures is the requirement to be a plane figure unnecessary?

In Problems 17–19, points M and N do not lie on $\angle ACB$. Does the line segment MN necessarily intersect $\angle ACB$ if:

17. Both M and N are interior points?
18. Both M and N are exterior points.
19. M is an interior point and N is an exterior point?
20. Describe the set $\overrightarrow{NM} \cup \overrightarrow{NP}$ if: (a) M, N, and P are collinear, that is, they lie on the same line, and N is between M and P; (b) M, N, P are collinear and P is between M and N; and (c) M, N, P are noncollinear.
21. Draw sketches to show that the intersection of two angles is (a) one point, (b) a line segment, (c) two points, (d) an empty set, (e) a ray, (f) three points, (g) a ray and a point not on the ray, (h) four points, and (i) a line segment and a point not on the line segment.
22. What is the measure in degrees of the angle formed by the hands of a clock at 6:30 o'clock?

23. POLYGONS AND THE CIRCLE

The student is probably familiar with the characterization of the illustrations in Figure 4.11 as polygons. The word "polygon" literally translated from the Greek means "many angles," although we usually think of a polygon as a "many-sided" figure.

Each of the polygons in Figure 4.11 has a specific name, derived from the number of line segments, or sides, it contains. These specific names will be discussed later. While a polygon has been defined as a "closed broken line in a plane," we shall develop a more adequate definition in what follows.

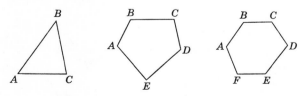

Figure 4.11

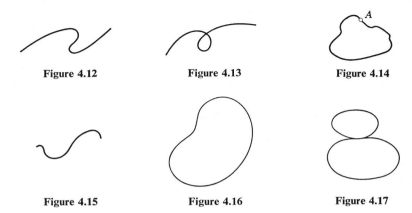

Figure 4.12   Figure 4.13   Figure 4.14

Figure 4.15   Figure 4.16   Figure 4.17

Simple Closed Curve

A *simple* curve is a curve that passes through any point it contains only once. Stated in another way, a simple curve is a curve that does not cross itself. Figure 4.12 is a simple curve, and Figure 4.13 is a nonsimple curve.

A closed curve, described rather than defined, is a curve that returns to its starting point, that is, one which has no endpoints. Figure 4.14 is a closed curve, while Figure 4.15 is an open curve. Of course, in a closed curve, any point, such as point *A*, may be considered to be the starting point.

A simple closed curve is then a curve that is both simple and closed. Figure 4.16 illustrates a simple closed curve, and Figure 4.17 illustrates a nonsimple closed curve.

Interior and Exterior of a Closed Curve

In Figure 4.18 the student would have little difficulty in describing point *A* as an interior point of the curve and point *B* as an exterior point. One of the important properties of a simple closed curve is that it separates a plane into

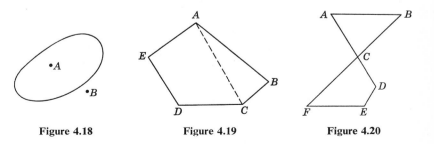

Figure 4.18   Figure 4.19   Figure 4.20

three disjoint sets of points: the curve itself, the interior of the curve, and the exterior of the curve. The interior of a closed curve is called a *region*. A region may be described as triangular, polygonal, and so forth, according to its boundary.

Definition of a Polygon

We are now in a position to define the term "polygon" and discuss some of its properties.

Definition 1. A polygon is the union of three or more line segments forming a simple closed curve. Each of the endpoints of a line segment is the intersection of two and only two segments.

Figure 4.19 illustrates a polygon. The line segments which form the boundary of the polygon are called *sides* of the polygon. The endpoints of each line segment are called *vertices*, and a line segment such as AC, which joins nonconsecutive vertices, is called a *diagonal* of the polygon. Even though the curve in Figure 4.20 is the union of three or more line segments, it is not a polygon because it is nonsimple.

Classification of Polygons

Polygons may be classified in a variety of ways, one of which groups all polygons as convex or concave. The ordinary use of the terms convex and concave seems to agree with Figures 4.21 and 4.22. Observing the two polygons suggests several possible definitions.

Each side of a polygon determines two half-planes. For example, FE in Figure 4.21 determines a half-plane which, together with the side FE, contains the entire polygon and another half-plane which does not contain the polygon. The polygon is convex if and only if the entire polygon lies in or on the edge of one of the half-planes determined by each side of the polygon. Note that this is true for each side of the polygon in Figure 4.21, but it is not true for sides BC and CD in Figure 4.22.

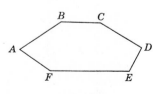

Figure 4.21

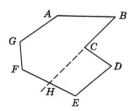

Figure 4.22

140   What are Some of the Fundamental Properties of Geometric Forms?

Figure 4.23

An alternate definition is based on the observation that in Figure 4.21 the line containing any side of the polygon does not again intersect the polygon, while in Figure 4.22 the line containing side *BC* intersects the polygon in point *H*. The definition of convex and concave polygons based on this property is stated as follows.

Definition 2. A polygon is convex if the line containing any side intersects the polygon only in the endpoints of that side. Polygons which are not convex are concave.

A second classification of polygons is based on the number of sides. Figure 4.23 illustrates several of these. The prefixes tri, quadra, penta, hexa, and octa are simply the Greek names for the number of sides of the polygon.

A polygon is classified as *regular* when all of its sides are equal in length and all of its angles have the same measure. Figure 4.24 illustrates a few regular polygons and their names.

Circles

While circles are not generally considered in the category of polygons, certain relations between them lead us to include them here.

Definition 3. A *circle* is a set of points in a plane, each of which is at a given distance from a point in the plane called the *center*.

Certain lines and line segments are basic to the study of the circle. Figure 4.25 illustrates these elements, and the definitions are as follows.

Definition 4. A radius is a segment one of whose endpoints is the center of the circle, the other endpoint being a point of the circle.

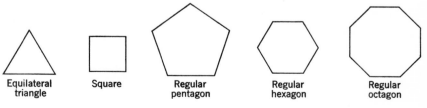

Figure 4.24

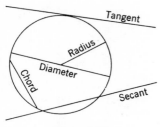
Figure 4.25

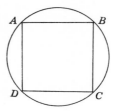
Figure 4.26

Definition 5. A diameter of a circle is a segment that contains the center and whose endpoints are points of the circle.

The measure of the diameter is clearly twice the measure of the radius. The diameter is a special case of a *chord* of the circle, defined as follows.

Definition 6. A chord of a circle is a segment both of whose endpoints are points of the circle.

Definition 7. A secant is a line containing a chord of the circle. It may also be defined as a line that intersects the circle in two distinct points.

Definition 8. A tangent to a circle is a line in the plane of the circle that intersects the circle in one and only one point, called the point of tangency.

Inscribed Polygons

When the vertices of a polygon are points of a circle, as in Figure 4.26, we say that the polygon is inscribed in the circle. We may also say that the circle is circumscribed about the polygon. Clearly, by choosing three or more points of a circle, we may always inscribe a polygon. A more interesting situation occurs when we wish to determine if a given polygon can be inscribed in a circle. Although any triangle may be inscribed in a circle, this is not always true of polygons of four or more sides. We shall only state here that it is always theoretically possible to inscribe a regular polygon in a circle; whether or not any other given polygons may be inscribed is a more difficult question.

The Circumference of a Circle

The circumference of a circle is the length, or measure, of the circle expressed in linear units. The problem of determining the circumference of a circle when the radius is known has a long history; today every student knows the familiar formula $C = 2\pi r$. Since π is an irrational number, it can never be expressed exactly as a common fraction or as a decimal fraction, but for most practical purposes the value $\pi = 3.1416$ is satisfactory.

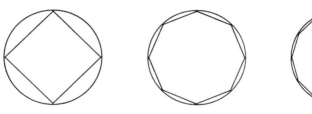

Figure 4.27

The concept of applying linear units of measure to a curved line is a fairly sophisticated mathematical idea. We shall introduce this idea in an intuitive but highly plausible manner. The inscribed regular polygons of Figure 4.27 suggest that as we increase the number of sides of the polygon the perimeters of the polygons more closely approach, but remain less, than the circumference of the circle. Stated in a different way, as the number of sides of an inscribed regular polygon increases, the difference between the perimeter of the polygon and the circumference of the circle decreases. It seems fairly clear that by continually increasing the number of sides of the polygon we can make this difference as small as we please. We state this mathematically by saying that the circumference is the *limit* of the perimeter of the regular inscribed polygon as the number of sides increases continually. This is a profound idea, basic to the theoretical development of calculus and all branches of higher mathematics.

That this conclusion is indeed only intuitive, that is, is in need of a more careful treatment, may be shown by the following consideration, noted by several modern mathematicians.[1] In Figure 4.28 the dotted path from A to B is made up of a series of equilateral triangles. It is clear that the length of the dotted path is twice the length of the straight-line path. Can we make them equal by making smaller triangles so that the path along the triangles is closer to the straight-line path? In Figures 4.29 and 4.30 we attempt to do this, yet a little reflection will show that the dotted path is still twice the straight-line path. In fact, no matter how close the two paths come to each other, the distances remain in a ratio of two to one. We must conclude, therefore, that in the case of the polygons inscribed in the circle something

Figure 4.28

[1] *What is Mathematics?*, Richard Courant and Herbert Robbins, Oxford, 1941, pp. 372–373.

Figure 4.29

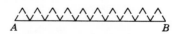
Figure 4.30

more than the closeness of the polygons to the circle or the chords to the arcs is needed to guarantee their equality. Although a rigorous treatment of this concept is beyond the scope of this book, it will be discussed further in Section 31.

EXERCISE 23

POLYGONS AND THE CIRCLE

1. Classify each of the curves of Figure 4.31 as simple, nonsimple, simple closed, or nonsimple closed.

(a)   (b)   (c)   (d)   (e)

Figure 4.31

2. Classify each of the polygons of Figure 4.32 as convex or concave.

 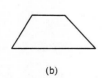
(a)   (b)   (c)   (d)

Figure 4.32

3. Show a drawing of a line containing (a) no point of a given simple closed curve, (b) one point of a given simple closed curve, and (c) two points of a given nonsimple closed curve.
4. Classify each point of Figure 4.33 as interior, exterior, or on the curve.

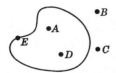

Figure 4.33

144 What are Some of the Fundamental Properties of Geometric Forms?

5. Describe the set of points in Figure 4.34 that satisfy each of the following conditions.
   (a) (interior of triangle ADB) ∩ $\overline{BD}$
   (b) (triangle ADB) ∩ (triangle BDC)
   (c) (triangle ABD) ∪ $\overleftrightarrow{BD}$
   (d) (exterior of triangle ABD) ∩ $\overline{BD}$
   (e) (exterior of triangle ABD) ∩ $\overrightarrow{BC}$
   (f) (exterior of triangle BDC) ∩ (triangle ABD)
   (g) (interior of triangle EFG) ∩ (exterior of circle O)

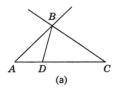 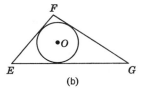

(a)　　　　　　　　　(b)

Figure 4.34

6. Sketch a broken line that is:
   (a) Simple but not closed.
   (b) Closed but not simple.
   (c) Simple, closed, and convex.
   (d) Simple, closed, and concave.

7. Sketch a number of circles each of which passes through the same two points. Describe the set of points composed of their centers.

8. Draw a circle through two points, A and B. Choose a third point, C, at random. In general, how many circles can be drawn through three points? How many circles can be drawn through three points on the same straight line?

9. Show that the rational number 355/113 is a better approximation to π than the more commonly used rational number 22/7.

10. Show by a sketch that a straight line may intersect a simple closed curve in more than one point.

11. Which of the regular polygons may be used for floor tiles?

In Problems 12–17, sketch a curve which is:

12. Simple and closed.
13. Nonsimple and closed.
14. Simple and not closed.
15. Nonsimple and not closed.
16. The boundary of a convex quadrilateral.
17. The boundary of a concave pentagon.

24. CONGRUENCE AND SIMILARITY

The concept of congruence is of prime importance in the study of geometry. Congruence is not only an important property of geometric figures, but it is

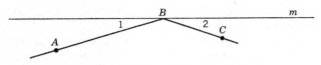

Figure 4.35

a means of deriving numerous other properties. An example is the proof that in Figure 4.35, where A and C are two fixed points in a plane and B is a point somewhere along line m in this plane, the path ABC is shortest when $\angle 1$ is congruent to $\angle 2$. This proof is easily established by means of congruent triangles and will be given later.

Congruence of Line Segments and Angles

In a previous section we discussed the measure of line segments and angles. These concepts are useful in stating the conditions of congruence for these geometric elements.

Two line segments are said to be congruent to each other if they have the same measure, regardless of the unit used. The symbol for the relation of congruence is $\cong$. Thus, in Figure 4.36, if the segments p and q have the same measure, they are congruent and we write $p \cong q$.

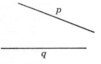

Figure 4.36

In Figure 4.37 suppose the segments $\overline{AB}$, $\overline{AC}$, $\overline{DE}$, and $\overline{DF}$ are congruent. Then the two angles BAC and EDF are congruent if $\overline{BC} \cong \overline{EF}$. The implication is that two angles are congruent if they have the same size, just as two line segments are congruent if they have the same length.

Congruence of Geometric Figures

The concept of congruence can now be extended to other sets of points such as polygons and circles. This extension is not simple; therefore, we shall first consider some definitions of congruence which have been stated by various writers. The first definition is fairly modern and may be found in textbooks today.

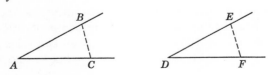

Figure 4.37

Definition 9. Two figures which have precisely the same size and shape are called congruent.

By this definition, the stamps in a sheet of postage stamps or the cards in a package of file cards would normally be considered congruent to each other. While the pairs of figures in Figure 4.38 seem to represent congruent polygons and circles, some difficulty may arise when it is required to prove that they are congruent, that is, that they have the same size and shape. In fact, the term "shape" would need to be defined in some manner before "sameness" of shape could be discussed.

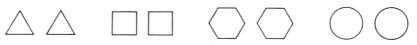

Figure 4.38

From a textbook dating back to the last century we find the following definition.[2]

Definition 10. If two figures can be made to coincide in all their parts, they are said to be congruent.

This is the basic idea of congruence, and one has but to stack the cards in a pack of file cards to illustrate it. While this definition avoids the use of the undefined term "shape," it also lacks the precise mathematical formulation needed to *prove* that any particular pair of figures is congruent. Early writers augmented this definition with the assumption that any figure may be moved from one place to another without altering its size and shape. This assumption permitted the figures to be moved (theoretically) to actually establish that they could be made to coincide.

We now state a modern definition of the general property of congruence.

Definition 11. Two figures are said to be congruent if and only if there exists a one-to-one correspondence between their points such that the segment determined by any two points of one figure is congruent to the segment determined by the corresponding points in the other figure.

This definition implies the coincidence of congruent figures and bases the congruence of these figures on the congruence of line segments.

Congruent Triangles

Since most geometric figures can be resolved into triangles, we state, but do not prove, three conditions for the congruence of triangles. In these statements, an angle is said to be *included* between two sides of a triangle if its rays contain the sides of the triangle, and a side is said to be included

---

[2] *Plane and Solid Geometry*, Wentworth, Smith, Ginn and Company, Boston, 1888, p. 26.

between two angles if it is on one of the rays of each of the angles. Thus, in Figure 4.39, $\angle A$ is included between the sides $\overline{AC}$ and $\overline{AB}$, and side $\overline{AB}$ is included between $\angle A$ and $\angle B$.

1. If two sides and the included angle of one triangle are congruent to two sides and the included angle of another triangle, then the triangles are congruent.
2. If two angles and the included side of one triangle are congruent to two angles and the included side of another triangle, then the triangles are congruent.
3. If each of the three sides of one triangle is congruent to the corresponding sides of another triangle, then the triangles are congruent.

The third condition for congruence is of great importance in engineering applications. It implies that a triangle is a "rigid figure," that is, the three sides determine its shape. A diagonal brace in the frame of Figure 4.40 applies this principle.

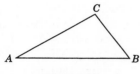

Figure 4.39

Figure 4.40

Returning to the problem suggested in the opening paragraph of this section, we are now in a position to prove the fact that the shortest path occurs when $\angle 1$ is congruent to $\angle 2$. It is a well-known law of optics that a ray of light following the path AB and striking a mirror at B will be reflected along the path BC such that $\angle 1$ is congruent to $\angle 2$. A ball following the same path AB and striking a wall at point B bounces back along the path BC. These facts are expressed by saying that "the angle of reflection is equal to the angle of incidence." What may be less familiar in the examples given is that of all possible paths from A to a point on line m and back to C, the path described, with $\angle 1$ congruent to $\angle 2$, is the shortest possible.

To prove informally that this is so, we reflect point A in Figure 4.41 to the other side of line m so that A' is a mirror image of A. Then $\overline{A'D}$ is congruent to $\overline{AD}$, $\overline{BD}$ is congruent to itself, and $\angle 4$ is congruent to $\angle 5$. We have at once from the first condition for congruent triangles that triangle BDA' is congruent to triangle BDA. It follows from the congruency of these triangles that $A'B$ is congruent to AB, and hence the path ABC is equal in length to the path $A'BC$. But this later path is shortest when $A'BC$ is a straight line. This

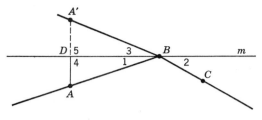

Figure 4.41

will occur when ∠3 is congruent to ∠2. But ∠3 is also congruent to ∠1, and hence *ABC* is the shortest path when ∠1 ≅ ∠2.

Similarity

It is fairly clear that, with the exception of the circle, the size and shape of a figure are independent of each other. Informally speaking, when two figures have the same shape but not necessarily the same size, they are said to be similar. An enlargement of a photograph is a good example of "sameness of shape" as is a mirror image of one's self. Trick mirrors in amusement centers distort the image, so that the sameness of shape is lost. Figure 4.42 illustrates several pairs of similar figures. It is to be noted that congruence is considered to be a special case of similarity. Of course, in proving two figures similar we cannot rely on the intuitively clear but mathematically vague notion of shape, just as we could not do so in establishing congruence.

To develop a working definition of similarity, it is necessary to recall the meaning of *ratio* and *proportion*. A ratio is used to compare two quantities of the same sort—the lengths of two line segments, the sizes of two angles, and so on. Since the comparison is made by utilizing the operation of division, the result always indicates either how many times as great one quantity is than the other, or what part one quantity is of the other. Thus, if the measure of a line segment *m* is 10 inches and that of a segment *n* is 2 inches, then the ratio of *m* to *n* is 5 and the ratio of *n* to *m* is 1/5.

A proportion is simply an equation stating that two ratios are equal. Since the ratio of a dime to a dollar (expressed in any common unit, as cents) is the same as the ratio of a millimeter to a centimeter (also expressed in a common

Figure 4.42

24. Congruence and Similarity   149

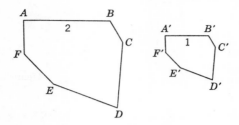

Figure 4.43

unit), then these two ratios can form a proportion of the form

$$\frac{\text{one dime}}{\text{one dollar}} = \frac{\text{one millimeter}}{\text{one centimeter}}$$

We return now to the discussion of similarity.

Definition 12. Two polygons with the same number of vertices (and hence sides) are similar if their corresponding angles are congruent and their corresponding sides are proportional.

Figure 4.43 illustrates two similar hexagons. Any letter and its prime, such as A and A', indicate corresponding angles, and any side such as $\overline{AB}$ has its corresponding side $\overline{A'B'}$. The ratios $\overline{AB}/\overline{A'B'}$, $\overline{BC}/\overline{B'C'}$, and $\overline{CD}/\overline{C'D'}$, are all equal to 2, and the corresponding angles are congruent.

Similar Triangles

The triangles in Figures 4.44 and 4.45 represent a special case of similarity of polygons. In Figure 4.44 all that is indicated is the proportionality of corresponding sides, and in Figure 4.45 all that is indicated is the congruence of corresponding angles. In each case the triangles are similar. It is possible to prove that when the polygon is a triangle the proportionality of corresponding sides *implies* the congruence of corresponding angles, and the congruence of corresponding angles *implies* the proportionality of corresponding sides.

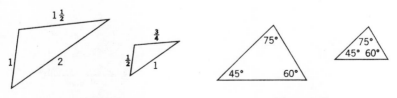

Figure 4.44   **Figure 4.45**

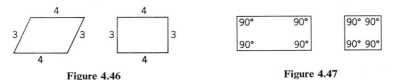

Figure 4.46   Figure 4.47

We note that the definition of similarity of polygons is redundant in the case of triangles, since more is said than is needed. In the case of polygons other than triangles, however, one condition does not imply the other. Thus, in Figure 4.46 the polygons are not similar even though the sides are proportional, and in Figure 4.47 the polygons are not similar even though the corresponding angles are congruent.

An Application of Similar Triangles

A useful application of the properties of similar triangles is the measurement of inaccessible heights and distances. Such measurement, where we cannot directly apply the measuring device, is called indirect measurement. Suppose that we wish to measure the height of the flagpole in Figure 4.48. Since it is quite impractical to apply a ruler or tape measure to the pole, we seek another method. If the rays of the sun strike the pole, a shadow will be cast along the ground; the length of this shadow can be easily measured. If we now place a yardstick vertically on the ground, as in Figure 4.49, we can also measure the length of its shadow. Since the angle of the sun's rays is the same in both cases, the corresponding angles in both figures are congruent and, therefore, the triangles are similar. We may then write the proportion

$$\frac{\overline{BC}}{\overline{B'C'}} = \frac{\overline{AC}}{\overline{A'C'}}$$

Since $\overline{BC}$ is the only unknown quantity in this proportion, we can calculate the height of the pole, that is, $\overline{BC}$.

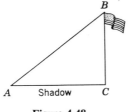

Figure 4.48

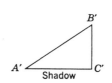

Figure 4.49

EXERCISE 24

Congruence and Similarity

In Problems 1–17 state whether the two given figures are necessarily congruent under the stated conditions.

1. Two circles with equal areas.
2. Two squares with equal areas.
3. Two rectangles with equal areas.
4. Two rectangles with equal areas and equal bases.
5. Two parallelograms with equal areas and equal bases.
6. Two parallelograms with equal areas and equal perimeters.
7. Two equilateral triangles with equal areas.
8. Two isosceles triangles with equal areas.
9. Two equilateral triangles with equal perimeters.
10. Two isosceles triangles with equal perimeters.
11. Two isosceles triangles with equal bases and equal perimeters.
12. Two isosceles triangles with equal areas and equal bases.
13. Two isosceles triangles with equal areas.
14. Two right triangles with equal areas and equal bases.
15. Two triangles with equal areas and equal bases.
16. Two hexagons inscribed in the same circle.
17. Two regular hexagons inscribed in the same circle.

In Figure 4.50, composed of congruent equilateral triangles, name all of the following figures.

18. The triangles congruent to triangle *ADE*.
19. The triangles similar to triangle *ADE*.

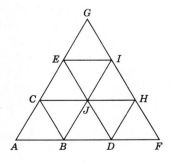

Figure 4.50

20. The trapezoids congruent to trapezoid *CBFG*.
21. The trapezoids congruent to trapezoid *EDFG*.
22. The trapezoids congruent to trapezoid *BCED*.
23. The trapezoids similar but not congruent to trapezoid *CBFG*.
24. In Figure 4.51, triangle *ABC* is isosceles, *D* is the midpoint of $\overline{AC}$, and *E* is the midpoint of $\overline{BC}$. Name two pairs of congruent triangles and state the condition for congruency which applies.

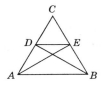

Figure 4.51

Write the converse, the inverse, and the contrapositive of each of the following statements and give the truth value of each. (Problems 25–28.)

25. If two triangles are congruent their corresponding angles are congruent.
26. If two triangles are similar their corresponding angles are congruent.
27. If two triangles are congruent their corresponding sides are congruent.
28. If two triangles are similar their corresponding sides are proportional.
29. A large floor is covered with black and white hexagonal tiles. Each white tile is surrounded by a single ring of black tiles. What is the approximate ratio of black tiles to white tiles?
30. A 5-foot stake casts a shadow 3 feet long at the same time a telephone pole casts a shadow 15 feet long. Find the height of the telephone pole.
31. In rectangle *ABCD*, $AB = DC = 24$, $AD = BC = 6$. In rectangle *EFGH*, $EF = HG = 40$ and $EH = FG = 10$. Are these rectangles similar?
32. If the word "parallelogram" is substituted for the word "rectangle" in Problem 31, are the two figures similar?
33. In triangle *ABC*, $AB = 7$, $BC = 4$, and $CA = 8$. In a similar triangle *DEF*, $DE = 6$. If vertex *A* corresponds to vertex *D*, vertex *B* corresponds to vertex *E*, and vertex *C* corresponds to vertex *F*, find the lengths of *DF* and *EF*.
34. Write the converse of the following statement: if two triangles are congruent, they are also similar. Is the converse true?
35. Write the contrapositive of the statement of Problem 34. What is the truth value of the contrapositive?
36. Write the inverse of the statement of Problem 34. Is it true?

25. GEOMETRIC FORMS IN SPACE

Thus far we have studied only plane geometry, that is, the geometry of figures in which all points and lines lie in the same plane. In space geometry, sometimes called solid geometry, we study figures whose points and lines are not limited to a single plane. In Section 22 we made brief mention of the term *surface* in describing a plane. *Surface* is one of the most difficult words to define, and any mathematically precise definition may prove meaningless to most students.[3] Fortunately the meaning of surface is not difficult to illustrate, nor to comprehend as an undefined term. A table top, the label on a can of food, or the skin of a human being are fairly good illustrations of a surface, although we would have to think of the label as well as the skin as having no thickness. The slogan of paint manufacturers, "Save the surface and you save all," is understood by anyone who reads it. We shall discuss several of the more common surfaces.

Cylindrical Surfaces

If a straight line moves along a plane curve d so that it passes successively through each point of the curve and is always parallel to a fixed line c, not in the plane of the curve, we say that it generates a *cylindrical surface*. In Figure 4.52 the curve d is called the *directrix*, and any of the straight lines is called an *element* of the surface. The plane in Figure 4.53 may be considered a special case of a cylindrical surface when the directrix is a straight line.

If, as in Figure 4.54, the directrix is a closed curve and the surface is bounded by two parallel planes which intersect it, the figure is called a *cylinder*. The portions of each plane bounded by the surface are called the

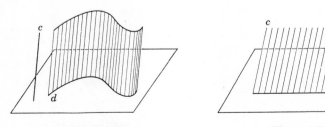

Figure 4.52   Figure 4.53

---

[3] An example is the definition: a surface in space is a set of points in one-to-one correspondence with a particular set of ordered number triples.

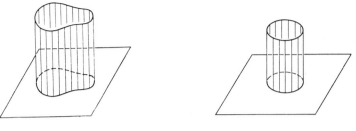

Figure 4.54   **Figure 4.55**

bases, and the cylindrical surface is called the *lateral surface*. The most common type of cylinder is the *right circular cylinder* (Figure 4.55) where the directrix is a circle and the elements of the cylindrical surface are perpendicular to the bases. The common "tin can" is perhaps the best-known example of a right circular cylinder.

Conical Surfaces

If, as in Figure 4.56, a straight line passes successively through each point of a directrix and through a point P not in the plane of the directrix, a *conical* surface is generated. Since the straight line need not terminate at P, there are two parts to the surface, called *nappes*, as in Figure 4.57. As in the cylindrical surface, each position of the straight line is called an *element* of the surface.

If, as in Figure 4.58, the directrix is a closed curve and the surface is bounded by the directrix and its interior points, the figure is called a *cone*. The most common type of cone is the *right circular cone* (Figure 4.59) where the directrix is a circle and the segment $\overline{PO}$, where O is the center of the circle, is perpendicular to the base of the cone. P is the *vertex* of the cone, and the conical surface is called the *lateral surface* of the cone.

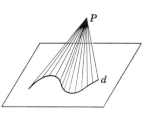 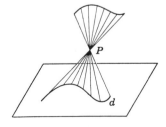

Figure 4.56   **Figure 4.57**

25. *Geometric Forms in Space*   155

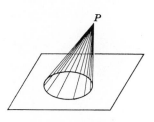

Figure 4.58

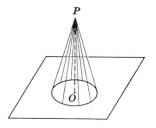

Figure 4.59

Pyramids

When, as in Figure 4.60, the directrix of a conical surface is a polygon, the cone is called a *pyramid*. Pyramids are named according to the polygons forming their bases, such as square pyramid, hexagonal pyramid, and so forth. *P* is the *vertex* of the pyramid, and the planes other than the base are called the *faces* of the pyramid. When the base is a regular polygon, the figure is called a *regular pyramid*.

Spherical Surfaces

The *spherical surface*, Figure 4.61, is defined as the set of all points in space at a fixed distance from a given point called the *center*. As in the case of the circle, this definition does not include the interior points of the surface, so that when we refer to a baseball or a billiard ball as a *sphere* we tacitly include the interior points as well.

Euler's Formula

An important formula, first explicitly stated by Descartes but named after the great mathematician, Leonhard Euler, who independently discovered it

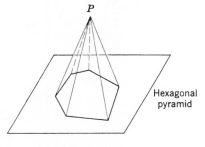

Figure 4.60

Figure 4.61

a century after Descartes,[4] applies to all polyhedrons. If V is the number of vertices, E the number of edges, and F the number of faces of any polyhedron, then Euler's formula states

$$V - E + F = 2$$

Thus, in a cube, $V = 8$, $E = 12$, and $F = 6$, and we have

$$8 - 12 + 6 = 2$$

Sections of a Surface

If a surface is cut by a plane, the resulting curve is called a *section* of the surface. A section is then the intersection of two sets of points, the points of the surface and the points of the cutting plane. It is not difficult to prove that all sections of a spherical surface are circles; any section passing through the center of the sphere is called a *great circle* of the sphere. Figure 4.62 illustrates sections of several surfaces.

The Conic Sections

The sections of a conical surface made by planes cutting the surface at various angles are of particular interest. These *conic sections* were known and studied as early as the third century B.C., and perhaps much earlier. Present-day analytic geometry studies these conic sections quite extensively on the basis of certain properties rather than as sections of a cone. Figure 4.63 illustrates the conic sections and their names. Many other surfaces not described here are best studied by considering various sections made by appropriate planes.

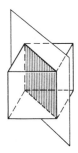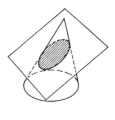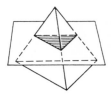

Figure 4.62

---

[4] See *An Introduction to the History of Mathematics*, Howard Eves, Holt, Rinehart & Winston, New York, 1958, p. 75.

25. Geometric Forms in Space   157

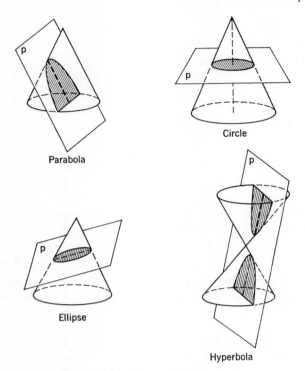

Figure 4.63

Polyhedrons

A polyhedron may be described as a closed surface bounded by planes, which are called the *faces*. Figure 4.64 shows a number of polyhedrons. The term polyhedron is frequently used to mean the entire solid figure, since it is a bounded surface. Of special interest are the five regular polyhedrons, so named because their faces are congruent polygons. The regular polyhedrons are often called the Platonic Bodies because they were extensively studied by pupils of the Greek philosopher Plato. The five regular polyhedrons are illustrated in Figure 4.65, their names indicating the number of faces. In

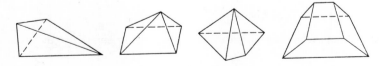

Figure 4.64

158   *What are Some of the Fundamental Properties of Geometric Forms?*

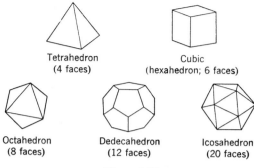

Figure 4.65

Problem 13 of Exercise 26 the student is asked to prove informally that no more than five regular polyhedrons are possible.

Prisms

A prism may be described as a closed surface such that two of its faces are polygons in parallel planes and the other faces are parallelograms. The two faces in parallel planes are called the *bases*, and the other faces are the *lateral faces* of the prism. The intersections of the lateral faces are called the *lateral edges* of the prism. The surface formed by the lateral faces is called a *prismatic surface*. Figure 4.66 illustrates several prisms, the most familiar being the *cube*, all of whose faces are congruent squares. It should be noted that a prism is a special case of a polyhedron.

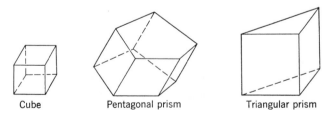

Figure 4.66

EXERCISE 25

Geometric Forms in Space

Describe and sketch each surface that is generated when point *P* moves in space according to the given conditions.

1. *P* moves at a fixed distance from a given plane.

2. *P* moves at a fixed distance from a given point.

3. *P* moves at a fixed distance from a given line.
4. *P* moves so that its distance from each of two fixed points remains the same.
5. *P* moves so that its distance from a given spherical surface remains the same.

Describe and sketch the surface generated in Problems 6–9.

6. The semicircle in Figure 4.67 revolves about the diameter *AB* as an axis.

Figure 4.67

7. A straight line moves along a given circle so that it passes successively through each point of the circle and is always perpendicular to the plane of the circle.
8. A straight line passes through a fixed point and passes successively through each point of a regular hexagon not in the plane of the fixed point.
9. A straight line passes successively through each point of a regular hexagon whose plane does not contain the line and is always parallel to a fixed line.
10. The section of a right, regular, prismatic surface by a plane will be a regular polygon if the cutting plane is ———.
11. The section of a pentagonal prismatic surface made by a plane parallel to one of the faces is ———.
12. Draw four cubes as in Figure 4.66. Then draw a section which is (a) a triangle, (b) a square, (c) the largest possible rectangle, (d) a hexagon.
13. Assume that the sum of the face angles of a polyhedron at any vertex must be less than 360°. Show that the only possible faces of a regular polyhedron could be equilateral triangles, squares, regular pentagons, and regular hexagons. (The face angles of a regular polyhedron are the angles between any two consecutive edges of the polyhedron.)
14. What is the least number of vertices a polyhedron may have?
15. What is the least number of faces a polyhedron may have?
16. What is the least number of edges a polyhedron may have?
17. Verify Euler's formula for the five regular polyhedrons.
18. Show that for a dodecahedron (12 faces), $F = 2E/5$.
19. Show that Euler's formula holds for a cross made up of five congruent cubes.
20. Each of the eight vertices of a cube is cut off so that the section is a triangle. Does Euler's formula hold?
21. Sketch a right prism whose lateral faces are squares. Is this prism always a cube?

26. SYMMETRY

An important property of geometric forms is *symmetry*, a concept not dealt with by Euclid in his *Elements*. Examples of symmetry are found abundantly in nature, in the works of man, and in man himself. The beauty of many figures is due to the property of symmetry, as exhibited by snowflakes, butterflies, and kaleidoscopes, and symmetry is often an essential ingredient of artistic design. Crystallography make extensive use of symmetry in the classification of crystals.

A glance at Figure 4.68, where all of the figures possess some form of symmetry, suggests that the attractiveness of each figure is due to the relation of the various parts to each other rather than to any property of the figure as a whole. In fact, the Encyclopedia Americana defines symmetry in nontechnical terms as "harmony or balance in proportion of parts as to the whole." The drawing of a typical kaleidoscope pattern shows that the arrangement of the colored bits of glass in any particular part of the pattern does not contribute to the beauty of the entire figure as much as the repetition of the pattern around the center of the total design. A turn of the kaleidoscope changes the pattern of the bits of glass, but the mirrors in the instrument retain the symmetry of the pattern about the center.

The definition quoted above, while expressing the notion of symmetry, lacks the precision required of a mathematical definition. To test any particular geometric form for symmetry requires more than the notion of "harmony and balance." We now state definitions of four types of symmetry found in geometric forms: *point* symmetry, *line* symmetry, *plane* symmetry, and *rotational* symmetry.

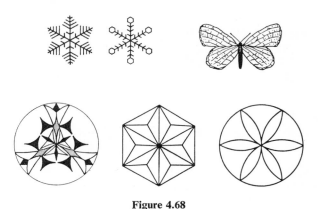

Figure 4.68

Point Symmetry

Definition 13. Two points, A and B, are symmetric with respect to a point C if C is the midpoint of $\overline{AB}$.

Figure 4.69 illustrates this definition. Point C is called the *center of symmetry*.

Definition 14. A geometric figure has point symmetry if every point of the figure has a symmetric point, with respect to the center of symmetry, on the figure.

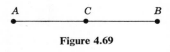

Figure 4.69

Figure 4.70 illustrates several examples of figures that have point symmetry.

Definition 15. Two geometric figures are symmetric with respect to a point if every point on one figure has a symmetric point, with respect to the center of symmetry, on the other.

Figure 4.71 illustrates several pairs of figures having point symmetry.

Line Symmetry

Definition 16. Two points, A and B are symmetric with respect to a line c if line c is the perpendicular bisector of $\overline{AB}$.

Figure 4.72 illustrates this definition.

As in the case of point symmetry, a geometric figure may have line symmetry and two figures may be symmetric with respect to a line, as illustrated in Figure 4.73. It is worth noting that a figure may have more than one line of symmetry, as in the case of the regular hexagon or the square, and that a figure may possess more than one kind of symmetry.

Plane Symmetry

Definition 17. Two points, A and B, are symmetric with respect to a plane MN if the plane is perpendicular to $\overline{AB}$ and bisects $\overline{AB}$.

Figure 4.74 illustrates this definition. Note that if plane MN were a mirror, point A would be the image of point B.

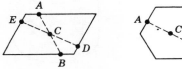

Figure 4.70

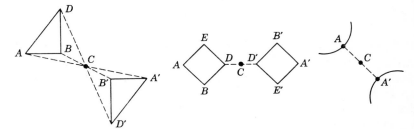

Figure 4.71

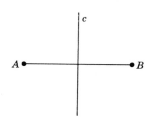

Figure 4.72

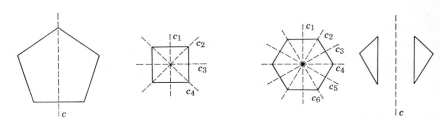

Figure 4.73

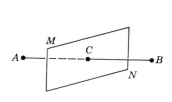

Figure 4.74

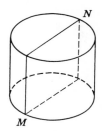

Figure 4.75

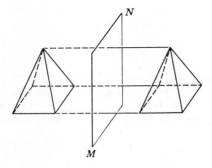

Figure 4.76

Definition 18. A surface is symmetric with respect to a plane if the plane cuts the surface into two parts, so that every point on one part has its symmetric point on the other.

Figure 4.75 illustrates this definition. As in the case of line symmetry, a surface may have more than one plane of symmetry. Figure 4.76 illustrates two surfaces symmetric with respect to a plane.

Rotational Symmetry

Definition 19. A figure has rotational symmetry if it can be made to coincide with itself by a rotation of $n°$, where $0° < n < 360°$.

Figure 4.77 illustrates several figures having rotational symmetry, with the value of n indicated.

Application of Symmetry to the Study of Curves and Surfaces

The principles of symmetry discussed in this section may be used to great advantage in the study of many curves and surfaces not usually considered in a course in plane or solid geometry. A number of these curves and surfaces are illustrated in Exercise 26. A more detailed study of their properties is an important part of analytic geometry.

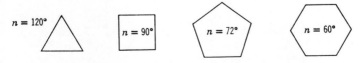

Figure 4.77

EXERCISE 26

SYMMETRY

1. For each of the given figures, draw the lines and the center of symmetry if they exist. Copy and complete the following table by checking in the first three rows the kinds of symmetry each figure possesses. In the last two rows give the number of lines of symmetry and the degrees of rotational symmetry.

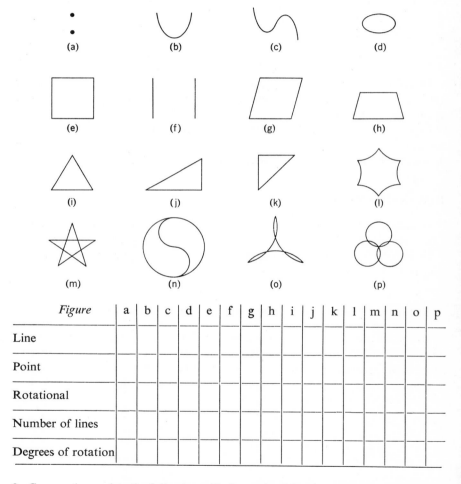

2. Copy and complete the following table for each of the given surfaces. Sketch each surface showing several planes of symmetry if more than one exists.

Surface	Planes of Symmetry	Number of Planes
(a) Right circular cylindrical surface		
(b) Right circular conical surface		
(c) Oblique circular conical surface		
(d) Rectangular prismatic surface		
(e) Cube		
(f) Regular tetrahedron		
(g) Sphere		

3. Sketch the figure that is symmetric to each of the given figures with respect to the dotted line as axis or with respect to the indicated point (P).

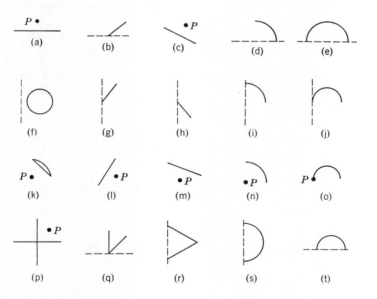

4. Each of the following sketches is part of a figure with rotational symmetry about the indicated point (O). There is 90° rotational symmetry for figures *a* to *f* and 120° rotational symmetry for figures *g* to *l*. Draw the entire figure.

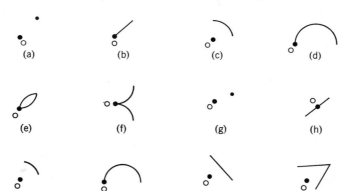

5. Discuss the symmetry of each of the following curves or surfaces.

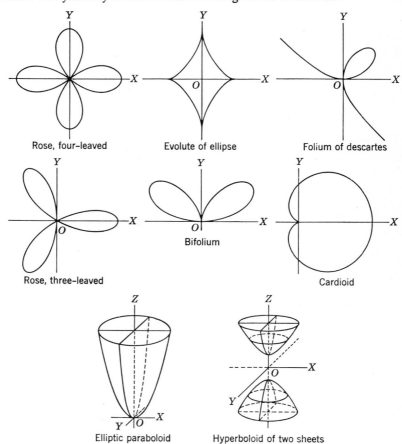

5

What is a Function?

27. THE FUNCTION CONCEPT

Although we have treated algebra and geometry as separate topics in the past two chapters, modern practice is to emphasize their similarities rather than their differences. A start in this direction has already been made; we may recall that the notion of *relation* occurs in the field of algebra (equality, inequality, and operations) and in the field of geometry (congruence and similarity). The notion of relation is a basic concept in establishing a unified approach to algebra and geometry. A special kind of relation, known as a *function*, has further importance in that it pervades much of the applied and theoretical aspects of modern mathematics. We will note, first, the distinction between a function and an ordinary relation.

Intuitively, a relation describes a correspondence between the elements of two sets. The correspondence will ordinarily take one of the following forms:
(a) one-to-one, as in the case of husbands and wives in a monogamous society (see Figure 5.1);
(b) many-to-one, as in the case of cities and states in the United States (see Figure 5.2); and
(c) one-to-many, as in the case of mothers and children (see Figure 5.3).
The first two examples are informal illustrations of a function; the third example illustrates a relation that is not a function. More explicitly, a

What is a Function?

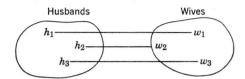

Figure 5.1

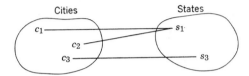

Figure 5.2

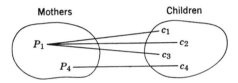

Figure 5.3

function is a relation that establishes a correspondence of *just one element of a second set* with each element of a first set. The correspondence may be one-to-one or many-to-one, the important criterion is that every object in a first set is associated with a single (not necessarily unique) object in a second set.

A little reflection will indicate that numerous important examples of functional relations occur naturally in the world about us. The speed with which a freely falling body strikes the ground is related to its initial height, the intensity of a sonic boom varies with the velocity of an aircraft, and the length of a metal span depends on its temperature. In each case there exists a correspondence between elements of two sets—one set consisting of values of a given variable and the other set consisting of single corresponding values of a related variable.

Definition of Function

Having introduced the notion of function in an informal way, we now consider a more useful and somewhat more sophisticated interpretation. Consider a variable x which represents any element of the set of real numbers, and a correspondence that assigns to each x an element $3x - 5$. As indicated

27. The Function Concept

in Figure 5.4, any element x_i (read x—sub i) in the first set will determine a corresponding element $3x_i - 5$ in the second set. (The symbol x_i is used to represent any one of the symbols $x_1, x_2, x_3, \ldots$). Specifically, as indicated in Figure 5.5, the correspondence assigns to the element 7 the element $(3 \cdot 7 - 5)$ or 16, to the element -2 the element -11, etc. For each element x in the first set, the corresponding element of the second set is called the *image of x*.

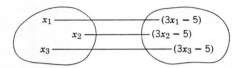

Figure 5.4

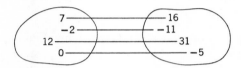

Figure 5.5

The following observations are significant.

1. The correspondence sets up a definite pairing of elements of the two sets. Note that when $x = 7$ its image is 16, but when $x = 16$ its image is not 7. Thus, the pair 7,16 is an ordered pair and, in the same way, every pairing of x and $3x - 5$ will generate an ordered pair of elements.
2. The value of $3x - 5$ depends on the value of x.
3. For each value of x there exists a single corresponding value of $3x - 5$.

These features are important mathematical characteristics of a function. The following is a formal definition.

Definition 1. A function is a correspondence between two sets that associates with each element of the first set a single element of the second set.

In the given illustration, the first set consists of all admissible values of x; it is called the *domain* of the function. The second set consists of all the images of x; it is called the *range* of the function. It is frequently convenient to use a single symbol to represent any element of the set of images. For example, if we assign the variable y to represent any element in the second set, then

$$y = 3x - 5$$

Since each value of x determines a corresponding value of y, we say that y depends on x. For this reason, y is called the *dependent variable* and x is called the *independent variable*.

What is a Function?

It is clear that y is the image of x and that the function is the set of ordered pairs resulting from the correspondence which matches each x with its image. In common practice, however, we frequently abbreviate this to the expression that *y is a function of x*.

The mathematical description of a function may take the form of an equation, a verbal statement, a tabular display of values, or a graph. The goal of a mathematical description is to provide useful information about the basic pattern of behavior of the related variables and to make possible the prediction of results for selected values of the variables.

EXAMPLE 1. The air temperature at a given weather observation post is a function which assigns to each time reading a corresponding temperature. The domain is the set of time readings and the range is the set of corresponding temperature readings. This function cannot generally be expressed in the form of an equation.

EXAMPLE 2. The area of a circular region is a function of its radius because for each value of the radius, r, the corresponding area, A, is determined by the equation $A = \pi r^2$. The domain of r is the set of all positive real numbers, and the range of A is the set of corresponding positive real numbers. The independent variable is r; the dependent variable is A.

EXAMPLE 3. The graph of Figure 5.6 is based on the mathematical probability that two persons will share the same day and month of birth. This graph defines a function whose domain is the set of numbers on the horizontal axis (the set of positive integers) and whose range is the set of numbers on the vertical axis (the set of rational numbers between 0 and 1). The probability is seen to increase sharply as the size of the group increases. In a random group of 35 people, the probability is about 0.8 or, expressed in another way, the chances are 4 out of 5 that two people in the group share the same birthday. Similarly, in a group of 25 people, the odds of this occurrence are approximately even.

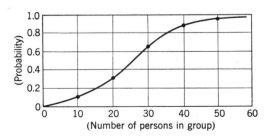

Figure 5.6 Probability of sharing a birthday.

EXAMPLE 4. The cost of printing an advertising circular is $350 for the first 10,000 copies plus $24.50 for each additional thousand. This verbal statement defines a function whose independent variable is the number of circulars printed and whose dependent variable is the cost of the printing.

Functional Notation

There are various conventions used to indicate that a function f transforms values of a variable x into its images. We will illustrate some of those in most common use. For this purpose let us consider again the function

$$y = 3x - 5$$

where set A is the domain of x and set B is the range of its images.

1. We can write

$$A \xrightarrow{f} B$$

which is read "the function from A to B," or we can express the specific rule of correspondence by writing

$$x \xrightarrow{f} 3x - 5$$

The images of particular values can be shown by the notation

$$3 \xrightarrow{f} 4 \qquad -8 \xrightarrow{f} -29 \qquad 0 \xrightarrow{f} -5$$

This notation readily conveys the idea that the function f determines a set of ordered pairs of elements.

2. Another common modern notation for the function is

$$f : A \to B$$
$$f : x \to 3x - 5$$

which is also read "the function from A to B" or "the function from x to $3x - 5$." The notation is sometimes read "the function f which maps A onto B," a terminology motivated by the fact that, in making a map of a geographic region, a correspondence is established between the geographic location and the map location of each point in the region. It is such a correspondence, of course, which is the essence of the function concept. We shall use the terms "function," "correspondence," and "mapping" interchangeably.

3. Perhaps the most usual notation, and the one we shall use in this book, is the use of the symbol $f(x)$, read "f of x," to represent the image of the element x. (It is important to note that this symbol does not mean f multiplied by x). Since the image of x is also represented by $3x - 5$, or y, we can write

$$y = f(x) = 3x - 5$$

An important feature of this notation is that since $f(x)$ represents the image of x, $f(a)$ represents the image of a, or the value that corresponds to a under the function f. For example, if
$$x = 7$$
then
$$f(x) = f(7) = 3 \cdot 7 - 5 = 16$$

EXAMPLE 5. If $f(x) = x^2 + 2x - 6$, find $f(5)$.
Solution: $f(5) = 5^2 + 2 \cdot 5 - 6 = 29$

EXAMPLE 6. If $f(t) = 8t + 40$, find $f(4) - f(3)$.
Solution: $f(4) = 8 \cdot 4 + 40 = 72$; $f(3) = 8 \cdot 3 + 40 = 64$
$$f(4) - f(3) = 72 - 64 = 8$$

EXAMPLE 7. Show that $f(a) = f(-a)$ if $f(x) = x^2 - 1$.
Solution: $f(a) = a^2 - 1$
$$f(-a) = (-a)^2 - 1 = a^2 - 1 = f(a)$$

EXERCISE 27

THE FUNCTION CONCEPT

A correspondence is formed between each element of set A and some element(s) of set B. Determine whether the resulting relation is or is not a function if:

1. $A = \{\text{persons}\}$, $B = \{\text{mothers}\}$
2. $A = \{\text{persons}\}$, $B = \{\text{ancestors}\}$
3. $A = \{\text{authors}\}$, $B = \{\text{books}\}$
4. $A = \{\text{real numbers}\}$, $B = \{\text{points on a number line}\}$
5. $A = \{\text{students}\}$, $B = \{\text{teachers}\}$
6. $A = \{\text{hour of sunrise in New York city}\}$, $B = \{\text{calendar date}\}$
7. $A = \{\text{common stocks listed on an exchange}\}$,
   $B = \{\text{closing prices on the exchange}\}$

Choose the one phrase that best completes each of the following statements.

8. The number of file cards which make up a 1 inch stack can be expressed as a function of (a) the area of each card, (b) the thickness of each card, (c) the weight of each card.
9. The speed with which a freely falling body strikes the ground can be expressed as a function of (a) the time it takes to fall, (b) its density, (c) its size.
10. The intensity of radiation emitted by a radioactive material can be expressed as a function of (a) its weight, (b) its shape, (c) its period of decomposition.

11. The distance of the earth from the sun is a variable which can be expressed as a function of (a) its speed in orbit, (b) the time of the year, (c) the change in the tides.
12. The area of a triangle with a fixed base can be expressed as a function of (a) its altitude, (b) its perimeter, (c) its base.

Describe the domain and range of each of the following functions if both the independent variable and its image are real numbers.

13. $f(x) = x + 1$
14. $f(x) = -x$
15. $f(x) = x^2$
16. $f(x) = \sqrt{9 - x}$
17. $f(x) = \sqrt{x^2 - 4}$
18. $f(x) = x^2 + 5$

For the function $f(x) = 7x - 8$, find the value of each of the following expressions.

19. $f(4)$
20. $f(5)$
21. $\dfrac{f(5) - f(4)}{5 - 4}$

22. $f(6)$
23. $f(11)$
24. $\dfrac{f(11) - f(6)}{11 - 6}$

25. $f(-3)$
26. $f(-5)$
27. $\dfrac{f(-3) - f(-5)}{(-3) - (-5)}$

For the function $f(x) = -3x + 4$, find the value of each of the following expressions (Problems 28–33).

28. $f(2)$
29. $f(6)$
30. $\dfrac{f(6) - f(2)}{6 - 2}$

31. $f(7)$
32. $f(12)$
33. $\dfrac{f(7) - f(12)}{7 - 12}$

34. Any function for which $f(x) = f(-x)$ is called an *even function*. Show that $f(x) = x^2$ is an even function, and give one other example of an even function.
35. Any function for which $f(x) = -f(-x)$ is called an *odd function*. Show that $f(x) = x^3$ is an odd function, and give one additional example of an odd function.

Each of the following formulas defines a function. (a) Identify the independent variable and its image. (b) Tell what each commonly represents. (c) Describe the domain and range of the function.

36. $C = 2\pi r$
37. $p = 4s$
38. $C = \tfrac{5}{9}(F - 32)$
39. $d = 30t$
40. $A = 50(1.06)^n$
41. $A = s^2$

Find the value of $f(3), f(0), f(-2)$ for each of the following functions (Problems 42–49).

42. $f(x) = 4x^2 - 8$
43. $f(x) = -x^2 - x$
44. $f(x) = x^3 - 1$
45. $f(x) = x^3 + x^2$
46. $f(x) = 2x^2 + 7x + 2$
47. $f(x) = -5x^2 + 2x - 1$
48. $f(x) = x - x^3$
49. $f(x) = 3 - x^2$

174   *What is a Function?*

50. The velocity in feet per second of a freely falling object is 32 times the time of motion expressed in seconds.

   (a) Express the relation between velocity, v, and time, t, by means of a formula.
   (b) Evaluate $f(5)$. What does it represent?
   (c) Evaluate $f(5) - f(4)$. What does this difference represent?

51. The blood pressure of a normal adult, measured in millimeters of mercury, is said to be 110 increased by half the person's age.

   (a) What formula describes the functional relation between the blood pressure p and the age n of an adult?
   (b) Evaluate $f(26)$. What does it represent?

52. The height of a triangle is 3 units greater than its base. If its base is x:

   (a) What formula describes the functional relation between the area of the triangle, A, and its base, x?
   (b) What is the domain of the function?
   (c) Evaluate $f(7)$. What does it represent?

53. Explain why each of the following graphs does or does not illustrate a function.

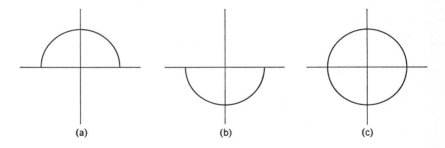

(a)   (b)   (c)

28. TABULAR REPRESENTATION

A table of values is a familiar method of depicting the correspondence between two sets of elements. Some common examples of tabular representation are found in tables of life expectancy, insurance premiums, tube characteristics, logarithms, trigonometric functions, annuities, compound interest, and powers and roots of numbers. Because of their obvious advantages of simplicity and ease of reference, tables are used extensively to display selected pairs of ordered values of a functional relation.

EXAMPLE 1. Consider the endless sequence of numbers $1, \frac{1}{2}, \frac{1}{4}, \frac{1}{8}, \frac{1}{16}, \ldots$, each obtained from its predecessor by multiplying by $\frac{1}{2}$. If S is the sum of the first x terms, then the functional relation between x and S can be presented in tabular form as follows:

x	1	2	3	4	5	6
S	1	1.5	1.750	1.875	1.9375	1.96875

EXAMPLE 2. The following table is a physical illustration of a functional relation between the time t required for one revolution of a planet about the sun and the average distance s of the planet from the sun, where the distance of the earth from the sun is taken as one unit. This relation is commonly known as Kepler's Law:

Planet	Mercury	Venus	Earth	Mars	Jupiter	Saturn	Uranus	Neptune	Pluto
s	0.387	0.723	1.00	1.52	5.20	9.54	19.2	30.1	39.5
t (yrs)	0.241	0.615	1.00	1.88	11.9	29.5	84.0	165	248

Constructing a Table of Values

The construction of a table of values is frequently based on observed measurements or collected statistics. For example, readings taken during controlled experiments can lead to a tabular representation between such related elements as:

(a) The price of a stock or bond and its yield in percent.
(b) The speed of an automobile and its stopping distance.
(c) The vapor pressure of a chemical and its temperature.
(d) The drill size and the tap size necessary to form a screw thread.

Such representation, based on empirical data, is essential when the function cannot be represented conveniently by an equation. For our purposes, however, a more important application of tabular representation occurs when a functional relation is defined by an equation or formula. Although the tabulation of ordered pairs may be merely an intermediate step in the process of graphing the relation, it is, nevertheless, a necessary and important procedure. The values of the variable which is regarded as independent are listed first, usually at equal intervals. The corresponding values of the dependent variable are then calculated by solving the equation or evaluating the formula.

EXAMPLE 3. Money invested at i percent per year will double in n years, where n is given approximately by the formula $n = 0.3 + 70/i$. Tabulate the relation for integral values of i between 1 and 10.

Solution:

i	1	2	3	4	5	6	7	8	9	10
n	70.3	35.3	23.6	17.5	14.3	12.0	10.3	9.1	8.1	7.3

EXAMPLE 4. If $f(x) = 4 + x^2$, construct a table showing values of $f(-3)$, $f(-1), f(1), f(3), f(5), f(7)$.

Solution: For $x = -3, f(x) = 4 + (-3)^2 = 4 + 9 = 13$. The remaining values are obtained in a similar manner, and the results are tabulated as follows:

x	-3	-1	1	3	5	7
$f(x)$	13	5	5	13	29	53

EXAMPLE 5. The value of n nickels and d dimes is 5 dollars. Tabulate the functional relation between d and n for $d = 1, 5, 9, 13, 17, 21, 25$.

Solution: The value of n nickels is $5n$ cents, and the value of d dimes is $10d$ cents. Therefore,

$$5n + 10d = 500$$

Substituting the given values of d into this equation, we obtain:

d	1	5	9	13	17	21	25
n	98	90	82	74	66	58	50

Expressing Tabular Representation by an Equation

Although the task of constructing a table of values from the algebraic description of a function is routine, the reverse process of determining an equation that satisfies the values listed in a table is complex and uncertain. Elaborate mathematical methods have been devised for this purpose, but we will confine ourselves to rather simple situations in which the results are intuitively obvious. Our objective is merely to demonstrate that an equation can frequently be associated with a given table of values.

Consider, for example, the following relation between x and y:

x	0	2	4	6	8	10	12
y	1	5	9	13	17	21	25

Looking only at the first ordered pair, that is, $y = 1$ when $x = 0$, we note that the relation between x and y could be any equation of the form

$$y = ax^n + 1$$

The second ordered pair $(2, 5)$ can be described by either the relation $y = x^2 + 1$ or $y = 2x + 1$. The third ordered pair $(4, 9)$ is satisfied by $y = 2x + 1$ but not by $x^2 + 1$. Similarly, each of the remaining pairs of corresponding values are satisfied by the relation $y = 2x + 1$ and, therefore, this equation can be said to describe the function depicted in the table. Of course, some individuals will note directly that each value of y is 1 more than twice the corresponding value of x and, therefore, that $y = 2x + 1$.

EXAMPLE 6. Given the following table, express y in the form $y = f(x)$:

x	-2	-1	0	1	2	3	4
y	4	1	0	1	4	9	16

Solution: Since each value of y is the square of the corresponding value of x,

$$y = x^2$$

EXAMPLE 7. Write an equation that describes the following function:

x	0	1	3	5	7	9
y	-1	0	8	24	48	80

Solution: Since each value of y is 1 less than the square of the corresponding value of x, $y = x^2 - 1$.

EXERCISE 28

TABULAR REPRESENTATION

Construct a table for each of the following functions to show the value of y corresponding to $x = -3, -2, -1, 0, 1, 2, 3, 4, 5$, respectively (Problems 1–9).

1. $y = 5x$
2. $y = 3x + 4$
3. $y = 2x - 7$
4. $y = -x - 3$
5. $y = -5x + 2$
6. $y = -4x - 1$
7. $y = x^2 + 3$
8. $y = x^2 + x - 1$
9. $y = -2x^2 + 4x$

10. (a) For which of the tables of Problems 1–9 is the interval between successive values of y constant?
    (b) For which is it not constant?
    (c) Is there any apparent relation between these findings and the nature of the equation which corresponds to the table?

178 What is a Function?

11. (a) Construct a table for each of the following functions for $x = 0, 1, 2, 3, 4, 5$:

$$y = 3x - 2$$
$$y = 3x + 6$$
$$y = 3x$$

(b) What can you say about the interval between successive values of y?

The length of one edge of a cube is represented by the independent variable x. Construct a table for the domain $1 \leq x \leq 10$ where x is an integer and its image is:

12. The area of one face.   **13.** The volume of the cube.
14. The total area of the six faces.
15. The length of the diagonal of one face.
16. The length of the diagonal of the cube.

For each of the following tables, write an equation that expresses y in terms of x.

17.

x	0	1	2	3	4	5
y	0	3	6	9	12	15

18.

x	0	1	2	3	4	5
y	1	4	7	10	13	16

19.

x	−3	−2	−1	0	1	2
y	3	2	1	0	−1	−2

20.

x	1	3	5	7	9
y	3	5	7	9	11

21.

x	−2	−1	0	1	2	3
y	−2	0	2	4	6	8

22.

x	−4	−2	0	2	4	6
y	−9	−5	−1	3	7	11

23.

x	0	1	2	3	4	5
y	0	2	8	18	32	50

24.

x	−2	−1	0	1	2	3	4
y	5	2	1	2	5	10	17

29. RECTANGULAR COORDINATE SYSTEM

A graph frequently provides an unusually clear and comprehensive view of a functional relation. Such representation is most commonly based on the Cartesian system of rectangular coordinates, which was introduced by René Descartes in 1637. The essence of the idea is that a correspondence can be established between ordered pairs of numbers and points in a plane, thereby making possible a transformation between geometric properties and algebraic (that is, numerical) properties. One important consequence is that the task of proving a geometric principle is cleverly converted to that of proving a corresponding principle in algebra.

29. Rectangular Coordinate System

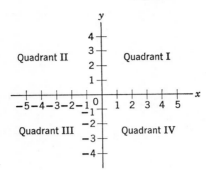

Figure 5.7

The system involves the following agreements.

1. Two real number lines, called the *coordinate axes*, are constructed perpendicular to each other. For convenience, one of the lines is drawn horizontally—it is called the *x-axis*, and the remaining line is called the *y-axis*.

2. The plane determined by the coordinate axes is called the *coordinate plane* or the *xy-plane*.

3. The axes divide the coordinate plane into four parts called *quadrants*. These are numbered counterclockwise I, II, III, IV, as shown in Figure 5.7.

4. The point of intersection of the axes is called the *origin*.

5. Ordinarily, the same unit of length is used on both axes, although in some cases it is convenient to do otherwise. Numbers associated with points on the *x*-axis are positive if the points are to the right of the origin, and they are negative if the points are to the left of the origin. Similarly, numbers associated with points on the *y*-axis are positive if the points are above the origin and negative if they are below the origin.

6. Each point in the coordinate plane determines an ordered pair of numbers and, conversely, each ordered pair of numbers determines a point in the plane. The ordered pair is written in the form (x, y) where the first component of the pair corresponds to the reading along the *x*-axis, and the second component corresponds to the reading along the *y*-axis. The first component is also called the *x-coordinate* or the *abscissa* of the point, and the second component is also called the *y-coordinate* or *ordinate* of the point.

7. Locating a point in the system is referred to as *plotting* the point, and the plotted point is called the graph of the ordered pair.

EXAMPLE 1. What ordered pairs correspond to points A and B in Figure 5.8?

Solution: The abscissa of point A is $+3$, and its ordinate is $+1$. Therefore, the ordered pair is $(3, 1)$. For point B, the abscissa is 0 and the ordinate is -2. Therefore its coordinates are $(0, -2)$.

180   What is a Function?

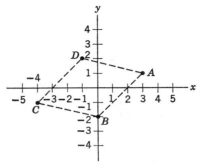

Figure 5.8

EXAMPLE 2. Plot the point $(-4, -1)$ in the xy-plane.

Solution: The point is 4 units to the left of the origin and 1 unit below the horizontal axis. Therefore, it is located at point C in Figure 5.8.

EXAMPLE 3. If points A, B, C (Figure 5.8) are three vertices of a parallelogram, what are the coordinates of the fourth vertex in the second quadrant?

Solution: There are three possible locations for a fourth vertex. In quadrant II the vertex is the intersection of lines parallel to AB and BC. Since point C is 4 units to the left of point B and 1 unit above it, the required vertex D will be 4 units to the left of, and 1 unit above, point A. Its coordinates are $(-1, 2)$.

EXERCISE 29

RECTANGULAR COORDINATE SYSTEM

1. Plot the points $(-4, -11)$ and $(5, 7)$. Draw a line through the two points. Which of the following points also lie on the line? $(-2, -7)$, $(-3, -6)$, $(-1, -4)$, $(0, -3)$, $(2, 3)$, $(3, 3)$, $(1, 1)$, $(4, 6)$.

Name the quadrant or quadrants for which each of the following statements is true (Problems 2–8).

2. The abscissa and the ordinate have unlike signs.
3. $x < 0$ 4. $y > 0$ 5. $x < 0, y < 0$
6. x and y have like signs.   7. The ordinate is negative.
8. The abscissa is positive, the ordinate is negative.

Plot each of the following sets of points. Then join the successive points of each set by straight lines and identify the geometric figure formed (Problems 9–13).

9. $(0, 0)$, $(6, -2)$, $(8, 4)$, $(2, 6)$
10. $(0, -3)$, $(-3, 3)$, $(1, 2)$, $(4, -4)$

11. $(-7, -2)$, $(14, 4)$, $(5, 7)$, $(-2, 5)$
12. $(-2, 6)$, $(1, 1)$, $(6, 4)$
13. $(-4, 1)$, $(-3, 5)$, $(9, 2)$, $(8, -2)$
14. (a) Draw a line through the points $(-4, 8)$ and $(6, -12)$.
    (b) List the coordinates of several other points on this line.
    (c) If y represents the ordinate of any one of these points and x represents the corresponding abscissa, is it true that in every case $y = -2x$?
    (d) Can you predict whether the point $(147, -294)$ would lie on the line?
    (e) Express the relation between the components of each ordered pair in the form of a verbal statement.
    (f) Does this relation appear to hold true even when x is not an integer?
15. (a) Plot the points $(-5, -3)$, $(-3, -1)$, $(-1, 1)$, $(3, 5)$, $(5, 7)$, $(7, 9)$.
    (b) Draw a line through the plotted points, and list the coordinates of several other points on the line.
    (c) Express the relations between the ordinate and abscissa of each point in the form of a verbal statement.
    (d) Express the relation between the ordinate and abscissa of each point in the form $y = f(x)$.
    (e) Will this line pass through the point $(-103, -102)$? Give a reason for your answer.
16. Referring to Figure 5.9, line segment AB joins $A(2, 1)$ and $B(17, 9)$.

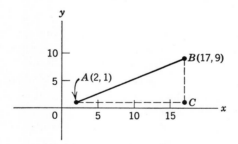

Figure 5.9

(a) What are the coordinates of point C if AC and BC are parallel to the respective axes?
(b) What is the length of the base of right triangle ABC?
(c) What is the height BC?
(d) Using the Pythagorean theorem, what is the length of the segment AB?

Using the Pythagorean theorem, as in Problem 16, determine the length of the line segment joining each of the following points (Problems 17–20).

17. $(0, 0)$ and $(8, 6)$
18. $(3, 1)$ and $(8, 13)$
19. $(-3, 2)$ and $(1, -1)$
20. (x_1, y_1) and (x_2, y_2)

182   What is a Function?

21. Prove that the triangle whose vertices are $(-5, 1)$, $(-2, 7)$ and $(4, 4)$ is isosceles.
22. Prove that the triangle whose vertices are $(-4, 3)$, $(-1, 4)$ and $(6, 5)$ is not isosceles.

Plot each of the following pairs of points, locate the midpoint of the line segment joining the pair of points, and determine the coordinates of each midpoint (Problems 23–28).

23. $(0, 0)$, $(4, 10)$ **24.** $(0, 0)$, $(-6, 2)$ **25.** $(-1, 3)$, $(3, 11)$
26. $(-2, -3)$, $(6, 5)$ **27.** $(-5, 4)$, $(-9, -2)$ **28.** $(-2, -1)$, $(-10, -9)$

Using the results of Problems 23–28:

29. How is the abscissa of the midpoint related to the abscissas of the given end points?
30. How is the ordinate of the midpoint related to the ordinates of the two given points?
31. Without plotting the points, what is the midpoint of the line segment joining the points $(1, 3)$ and $(19, -17)$?
32. What is the midpoint of the line segment joining the points (x_1, y_1) and (x_2, y_2)?

30. GRAPHIC REPRESENTATION

If both the domain and range of a function are real numbers, then the function may be represented by a graph in the Cartesian coordinate system. We define the graph of a function as follows.

Definition 2. *The graph of a function whose ordered pairs are (x, y), where x and y are real numbers, is the set of points in the xy-plane whose coordinates are the given pairs.*

The set of points may lie on a straight line, in which case we can draw any convenient portion of the line, or the set of points may lie on a curve, and then we usually approximate the graph by drawing a smooth curve through the plotted points.

Point by Point Plotting

We consider first a rather direct approach to graphing—simply selecting a succession of points to determine the general characteristics of the curve, supplemented by as many intermediate points as are needed to refine specific portions of the curve.

EXAMPLE 1. Graph the function $y = x^3$.

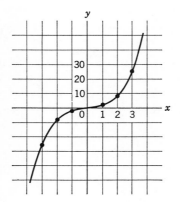

Figure 5.10

Solution: Choosing values of x and computing the corresponding values of y, we have:

x	0	1	2	3	-1	-2
y	0	1	8	27	-1	-8

Discussion. (See Figure 5.10.) The curve changes very rapidly in the first quadrant as x increases and in the third quadrant as x decreases. Of particular interest is its behavior near the origin. Fractional values of x will indicate that the ordinate changes very slowly in comparison to the change in the abscissa as the curve approaches the origin.

EXAMPLE 2. Draw a graph of the set of ordered pairs which satisfies the relation

$$3x - 2y = 12$$

Solution: Choosing values of x in the interval $0 \leq x \leq 10$, we plot the points determined in the following table:

x	0	2	4	6	8	10
y	-6	-3	0	3	6	9

Discussion. (See Figure 5.11.) The graph appears to be a straight line passing through the third, fourth, and first quadrants.

EXAMPLE 3. Draw the graph of the function $f(x) = x^2 - 9$.

184 *What is a Function?*

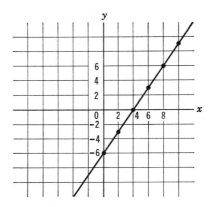

Figure 5.11

Solution: The following table lists a sufficient sequence of ordered pairs to determine the general nature of the curve:

x	0	1	2	3	4	5	-1	-2	-3	-4
$y = f(x)$	-9	-8	-5	0	7	16	-8	-5	0	7

Discussion: (See Figure 5.12.) The domain of the function is the set of real numbers. The range of the function appears to be limited to real numbers equal to or greater than -9. To determine whether -9 is indeed the lower limit of the range, it is possible to investigate the value of y for values of x as close to $x = 0$ as we please.

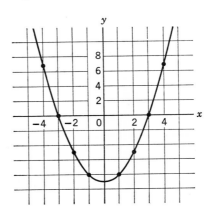

Figure 5.12

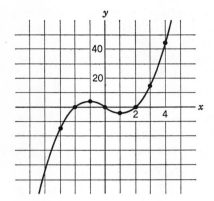

Figure 5.13

EXAMPLE 4. Graph the function $y = x^3 - 4x$.

Solution: Choosing zero and positive values for x and solving the resulting equation for y, we have:

x	0	1	2	3	4
y	0	-3	0	15	48

It is evident that as x increases, y will continue to increase very rapidly. Choosing negative values of x, we have

x	-1	-2	-3
y	3	0	-15

Discussion. (See Figure 5.13.) The domain and range of the function have no restrictions in the set of real numbers. The curve crosses the x-axis in three distinct points and the y-axis at a single point—the origin.

General Procedures for Graphing

The selection of plotted points in determining the graph of a function need not be entirely random—it is possible to utilize some general procedures which will not only reduce the number of points needed but will determine the fundamental characteristics of the curve with greater accuracy and clarity. We introduce several such procedures in the expectation that the student will make increasingly greater use of them as he increases his ability to handle the concomitant mathematical operations.

1. Intercepts. The intercepts are the points at which the curve crosses the coordinate axes. The y-intercept is found by setting $x = 0$ in the relation $y = f(x)$ and solving for y. That this is usually a simple matter can be ascertained by verifying the y-intercept in Examples 1 to 4. Similarly, the x-intercept is found by setting $y = 0$ and solving for x. This may, at times, lead to a second degree equation as in Example 3 or a third degree equation as in Example 4.

2. Symmetry. The points (x, y) and $(-x, y)$ are symmetric with respect to the y-axis. Therefore, any function for which $f(x) = f(-x)$ will result in a curve which is symmetric with respect to the y-axis. We note that for Example 3,
$$f(x) = x^2 - 9$$
$$f(-x) = (-x)^2 - 9 = x^2 - 9$$

Therefore, the curve is symmetric with respect to the y-axis. Knowing this fact, we could merely plot the curve for $x \geq 0$ and draw the mirror image of the curve for $x < 0$. The symmetry is evident in the graph for Example 3.

Similarly the points (x, y) and $(-x, -y)$ are symmetric with respect to the origin. Therefore, if the replacement of (x, y) with $(-x, -y)$ leaves the given equation unchanged, the graph will be symmetric with respect to the origin. In Example 4, $y = x^3 - 4x$. If we replace x by $-x$ and y by $-y$, we have
$$-y = (-x)^3 - 4(-x)$$
$$-y = -x^3 + 4x$$
$$y = x^3 - 4x$$

Since the equation is unchanged by the substitution, the resulting curve, as illustrated in Figure 5.13, is symmetric with respect to the origin. So, too, is the curve discussed in Example 1.

3. Domain and Range. A knowledge of any limitations on the domain and range of a function may be useful because they identify portions of the coordinate plane to which the graph is confined. We have seen in Example 3 that the range of the function was limited to $y \geq -9$. If we write the relation of Example 3 in the form $y = x^2 - 9$ and solve for x^2, we have
$$x^2 = y + 9$$

A fundamental algebraic principle pertains here: expressions equal to a perfect square cannot be negative. Therefore, solving the inequality,
$$y + 9 \geq 0$$
$$y \geq -9$$

Other common restrictions on the domain and range of a function will be considered in a subsequent chapter dealing with quadratic expressions.

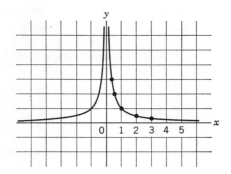

Figure 5.14

EXAMPLE 5. Discuss the graph of the function $y = 1/x^2$.

Intercepts. Since division by zero is undefined, there is no y-intercept. Similarly, there is no x-intercept.

Symmetry. The curve will be symmetric with respect to the y-axis, since $f(x) = f(-x)$.

Domain. There is no limitation on x, except that $x \neq 0$.

Range. Since $x^2 = 1/y$, y must be positive. Therefore, the curve exists only in the first and second quadrants.

We now plot a few points in the first quadrant, noting that as x increases y decreases.

x	$\frac{1}{2}$	1	2	3	$\cdots$	10
y	4	1	$\frac{1}{4}$	$\frac{1}{9}$	$\cdots$	$\frac{1}{100}$

Using the principle of symmetry, the second branch of the curve is drawn in the second quadrant. The curve is illustrated in Figure 5.14.

EXERCISE 30

GRAPHIC REPRESENTATION

Draw the graph of the functions defined by each of the following equations.

1. $y = 3x - 2$
2. $y = 2x + 7$
3. $y = -x - 2$
4. $y = -\frac{1}{2}x + 1$
5. $3x - 2y = 6$
6. $x + 5y = 15$
7. $y = -x^3$
8. $y = \frac{1}{2}x^3$
9. $y = 12/x$
10. $y = -(6/x)$
11. $y = x^3 - x^2$
12. $y = x^3 + 3x^2 - 2x$

13. $y = \frac{1}{2}x^4$
14. $y = -x^4$
15. $y = x^2 - 7x + 10$
16. $y = 4/x^2$
17. $y = 2x^2 - 8x$
18. $y = -x^2 + 2x - 1$

Give the coordinates of the point that is symmetric to the given point with respect to (a) the y-axis, (b) the x-axis, (c) the origin.

19. $(5, 3)$
20. $(-1, 8)$
21. $(4, -7)$
22. $(-2, -6)$
23. $(7, 0)$
24. $(0, -9)$
25. (a, b)
26. $(a, -b)$
27. $(-a, b)$

Without graphing, discuss the symmetry (if any) of the curve that corresponds to each of the following functions.

28. $y = x^2$
29. $y = -x$
30. $y = x^3 - x$
31. $y = x^4 + 3x^2$
32. $y = 3x + 1$
33. $y = x^2 + 4$

Without plotting, determine the intercepts (if any) of the curve that corresponds to each of the following functions.

34. $2x + 7y = 28$
35. $x - 3y = 18$
36. $y = (x - 2)(x - 5)$
37. $y = (x - 1)(x - 2)(x + 3)$
38. $x^2 + y = 4$
39. $xy = 5$

Sketch and discuss the graph of each of the following functions with respect to intercepts, symmetry, domain, and range.

40. $y = 2x$
41. $y = 2x^2$
42. $y = (x - 2)(x + 2)$
43. $y = (x - 1)(x + 2)$
44. $y = (x^2 - 9)(x^2 - 1)$
45. $x^2 y = 9$

31. FUNCTIONS AND CHANGE

Everything in the world is subject to change, although at times the change may be very subtle—the buildup of moisture in the air, the expansion of a bridge in the sunlight, the drop-off of voltage in an electric line, or the decrease in radioactivity of a substance over the centuries. By way of contrast, other changes may be very marked—the explosive force of a chemical reaction, the rapid acceleration of an object plunging through the earth's gravitational field, or the increased intensity of sound as a jet engine is warmed up. The inevitability of change and its scientific importance has created a need for a mathematical tool to deal with its measurement. We introduce now and will expand in the next several chapters the use of functions in providing a basis for the mathematical analysis of change.

Average Rate of Change

Let us consider a function $y = f(x)$ satisfied by the ordered pairs indicated in the following table:

x	3	7
y	12	30

It appears that in the given interval y increases more rapidly than x, specifically, y increases 18 units while x increases 4 units. The ratio of the change in y compared to the change in x is called the *average rate of change* of the function in the given interval. For this example, the average rate of change is $18/4$ or $4\frac{1}{2}$. This means that y tends to increase $4\frac{1}{2}$ units for each unit increase in x over the given interval. Note that if the change would occur at this rate it would lead to the following intermediate values in the interval:

x	3	4	5	6	7
y	12	$16\frac{1}{2}$	21	$25\frac{1}{2}$	30

We cannot be certain that the function does, in fact, behave in this manner throughout the interval, but it is a reasonable and useful conjecture in the absence of any contrary information.

Consider now a general interval for the function $y = f(x)$ as indicated in the following table:

x	x	$(x + h)$
y	$f(x)$	$f(x + h)$

The change in y is $f(x + h) - f(x)$, and the corresponding change in x is $(x + h) - x$ or h. The average rate of change of y with respect to x in the interval is given by the relation:

$$\text{Average rate of change} \Big|_x^{x+h} = \frac{f(x + h) - f(x)}{h}$$

The left member of this relation is read, "The average rate of change in the interval x to $x + h$." The application of this relation to specific functions is illustrated in the following examples.

EXAMPLE 1. Find the average rate of change of the function $y = 4x - 3$ in the interval $x = 2$ to $x = 9$.

Solution:

$$\text{Average rate of change} \Big|_2^9 = \frac{f(9) - f(2)}{7} = \frac{33 - 5}{7} = 4$$

What is a Function?

Therefore, for the function $y = 4x - 3$, y increases 4 units per unit increase in x in the interval $x = 2$ to $x = 9$.

EXAMPLE 2. If $y = -3x^2 + x$, what is the average rate of change of the function as x changes from 1 to 3?
Solution:

$$\text{Average rate of change}\Big|_1^3 = \frac{f(3) - f(1)}{2} = \frac{-24 + 2}{2} = \frac{-22}{2} = -11$$

Therefore y is decreasing 11 units per unit increase in x in the given interval.

Rate of Change Units

Since the rate of change is a ratio, the units in which it is measured will be of the form: units of measure of the dependent variable per unit of measure of the independent variable. For example, if y is measured in feet and x is measured in seconds, then the change is measured in feet per second—a common unit of velocity. On the other hand, if y represents velocity and x represents time, then the ratio represents the change of velocity with respect to time—commonly identified as acceleration. The actual units of acceleration may be miles per hour per minute, or feet per second per second, and so on. Rate of change units need not be confined to those dealing with motion. Thus, if y is measured in pounds and x in cubic inches, then the rate of change is concerned with pounds per cubic inch—a measure of density. Or y can represent a sum of money and x a time factor, in which case the rate of change is measured in dollars per unit time—a measure of return on investment. Other rate of change units and their common interpretations will suggest themselves as the student forms the ratio of measures of the related variables.

EXAMPLE 3. Water is flowing out of a tank in such a way that the number of gallons left after t minutes is given by $G = 1280 - 4t$. What is the average rate of flow during the first 30 minutes?
Solution:

$$\text{Average rate of change}\Big|_0^{30} = \frac{f(30) - f(0)}{30} = \frac{1160 - 1280}{30}$$

$$= \frac{-120}{30} = -4 \text{ gal/min}$$

Therefore, the volume (in gallons) is decreasing at an average rate of 4 gallons per minute in the given interval.

Instantaneous Rate of Change

Actually, the measure of the average rate of change is a routine and almost trivial mathematical procedure. Of far greater significance and challenge is the measure at a given instant of a change which is presumed to be continuous. Such a measure is called the *instantaneous rate of change*. Mathematicians have devised a remarkably ingenious method for determining the instantaneous rate of change in any situation which can be expressed in terms of an equation. The essence of the procedure is this.

1. Determine the change in the dependent variable which results from the change in the independent variable over a small interval.

2. Form the ratio of these changes in the form:

$$\frac{\text{change in dependent variable}}{\text{corresponding change in independent variable}}$$

This ratio, of course, is the average rate of change over the selected interval.

3. Let these changes diminish until they *approach* zero. The average rate of change over such a diminishing interval will then approach the instantaneous rate of change.

The key to the procedure is to find the value of the ratio of the changes as the interval shrinks toward zero. This value is called a *limit*—a mathematical concept we encountered in geometric form in Section 23. We reintroduce it now from an algebraic point of view.

Limits

Let us contrast the behavior of the two selected functions

$$f_1(x) = \frac{x^2 - 4}{x - 2} \qquad f_2(x) = \frac{x - 2}{x^3 - 3x^2 + 4}$$

for values of x in the vicinity $x = 2$. Note that when $x = 2$ we cannot evaluate either function, since division by zero is not defined. We examine the behavior of each function as x *approaches* the value 2:

Table 5.1

x	1	1.5	1.9	1.99	1.999	...
$f_1(x)$	3	3.5	3.9	3.99	3.999	...

Table 5.2

x	1	1.5	1.9	1.99	...
$f_2(x)$	−0.5	−0.8	−3.45	−33.4	...

In the case of the first function, it appears that as x approaches 2 the ratio $f_1(x)$ approaches 4. We say that the limit of $(x^2 - 4)/(x - 2)$ as x approaches 2 is 4, and write this as

$$\operatorname*{Lim}_{x \to 2} \frac{x^2 - 4}{x - 2} = 4$$

Note that at no time do we claim that the ratio is 4, merely that the *limit* is 4. By this we mean that the ratio $f_1(x)$ can be made as close to the value 4 as we please.

The ratio $f_2(x)$ is quite another matter. As x approaches 2, $f_2(x)$ does not appear to approach any limit. We say that as x approaches 2, the function $f_2(x)$ has no limit. We now define the concept of limit as follows.

Definition 3. Let $f(x)$ be a function of x, and let n be some constant. If there is a number L such that we can make the value of $f(x)$ as close to L as may be desired by choosing x sufficiently close to n, then the limit of $f(x)$, as x approaches n, is L. In symbols,

$$\operatorname*{Lim}_{x \to n} f(x) = L$$

EXAMPLE 4. Evaluate

$$\operatorname*{Lim}_{x \to 1} \frac{1 - x}{1 - x^2}$$

Solution: By direct evaluation we have:

x	0	0.5	0.9	0.99	0.999	...
$f(x)$	1	0.67	0.53	0.503	0.5003	...

It appears that as x approaches 1, the given ratio approaches 1/2. This conclusion can be given further support by letting x approach 1 from values $x > 1$. A somewhat more elegant method to determine this limit is to observe that for all values of x except $x = 1$,

$$\frac{1 - x}{1 - x^2} = \frac{1 - x}{(1 - x)(1 + x)} = \frac{1}{1 + x}$$

Therefore

$$\operatorname*{Lim}_{x \to 1} \frac{1 - x}{1 - x^2} = \operatorname*{Lim}_{x \to 1} \frac{1}{1 + x} = \frac{1}{1 + 1} = \frac{1}{2}$$

31. Functions and Change

We now summarize the two important ideas we have introduced concerning the rate of change of functions.

1. *Average rate of change.* For any function $f(x)$, the average rate of change in the interval x to $(x + h)$ is given by the expression

$$\frac{f(x + h) - f(x)}{h}$$

2. *Instantaneous rate of change.* For any function $f(x)$, the instantaneous rate of change at a particular value of x is the limit of the average rate of change as the interval x to $(x + h)$ approaches zero. It can be evaluated by the expression

$$\lim_{h \to 0} \frac{f(x + h) - f(x)}{h}$$

In the several chapters which follow we will further develop these concepts as we illustrate their application to specific functions.

EXERCISE 31

Functions and Change

In Problems 1–10, identify the units of measure of the change of y with respect to x if:

1. y represents distance in miles and x represents time in hours.
2. y represents temperature in degrees centigrade and x represents time in minutes.
3. y represents speed in miles per hour and x represents time in seconds.
4. y represents area in square miles and x represents time in hours.
5. y represents intensity of sound in decibels and x represents distance in feet.
6. y represents the number of bacteria in a culture and x represents time in hours.
7. y represents the length of a coiled spring in centimeters and x represents weight in grams.
8. y represents the area of a wound in square centimeters and x represents time in days.
9. y represents distance in miles and x represents gasoline consumption in gallons.
10. y represents volume of gas in cubic inches and x represents pressure in pounds.
11. Given the function defined by the table:

x	1	3	5	7	9
y	4	1	-2	-5	-8

(a) Find the average rate of change for each interval in the table.

(b) What interpretation can you give the sign associated with this average rate of change?
(c) What value of y could reasonably be associated with $x = 2$?

12. Using the function $y = x^2 + x$:
    (a) Complete a table for $x = 1, 3, 5, 7, 9$.
    (b) Find the average rate of change for each interval in the table.
    (c) Is the average rate of change the same in each interval or does it vary?
    (d) If we were to assume that the average rate of change is constant throughout a given interval, what tabular value of y could reasonably be associated with $x = 2$?
    (e) Using the given equation, what value of y is associated with $x = 2$?

In Problems 13–18, evaluate $\dfrac{f(x+h) - f(x)}{h}$ for $x = 3, h = 2$ if:

13. $f(x) = 2x + 7$
14. $f(x) = x^2 - 5$
15. $f(x) = 3x^2 + x + 2$
16. $f(x) = 1 - x$
17. $f(x) = -x^2 + 2x$
18. $f(x) = x^3 - 5x$

19. The total accumulation of an investment of $1000 at 5% simple interest is given by $A = 1000 + 50t$, where A is measured in dollars and t in years. Find the average rate of change during the first three years.

20. If $1000 is invested at 5% compounded annually, the total accumulation after t years is given by $A = 1000(1.05)^t$. Find the average rate of change during the first three years.

21. A free-wheeling railroad car rolls down an incline a distance of s feet in t seconds, where $s = 3t^2$. Find the average rate of change in the interval:
    (a) $t = 4$ to $t = 5$
    (b) $t = 4$ to $t = 4.5$
    (c) $t = 4$ to $t = 4.1$
    (d) $t = 4$ to $t = 4.01$

22. (a) Based on the results of Problem 21, does the average rate of change appear to approach a limit as the interval approaches zero?
    (b) If so, what is the limit and what does it represent?

23. Consider the function $y = x^2 + 1$.
    (a) Determine the average rate of change in the interval $x = 1$ to $x = 1 + h$.
    (b) Using successively smaller replacements for h, what numerical limit does the average rate of change approach as h approaches zero? This is the instantaneous rate of change at $x = 1$.
    (c) Determine the average rate of change in the interval $x = 1$ to $x = 1 - h$.
    (d) What limit does this average rate of change approach as h approaches zero?

Evaluate the limit (if it exists) of each of the following by examining the behavior of the function in the vicinity of the given value of x.

24. $\displaystyle\lim_{x \to 0} \dfrac{x}{x^2 - x}$
25. $\displaystyle\lim_{x \to 1} \dfrac{1}{x - 1}$
26. $\displaystyle\lim_{x \to 1} \dfrac{x - x^2}{1 - x^2}$
27. $\displaystyle\lim_{x \to -3} \dfrac{x^2 - 9}{x^2 + 6x + 9}$

6

What is a Linear Function?

32. THE LINEAR FUNCTION

An important reason for the study of functions is to recognize the common elements of relations that have similar properties. For example, the following familiar formulas can be viewed as particular forms of a single, more general function.

$F = 9/5\ C + 32$ The Fahrenheit reading corresponding to a given centigrade reading
$E = I \cdot R$ The relation between voltage, amperage, and resistance in an electrical circuit
$d = r \cdot t$ The distance covered at a rate r in a time t
$A = p + prt$ The simple interest relation
$C = 2\pi r$ The relation between the circumference and radius of any circle

The function that is exemplified here is called a *linear function*, and its general form is

$$y = f(x) = mx + b$$

where m and b are constants. The formula $F = 9/5\ C + 32$ is clearly of this form, with $m = 9/5$ and $b = 32$. So, also, is the formula $E = I \cdot R$ if we take

What is a Linear Function?

$m = 1$ and $b = 0$. Similarly, each of the other formulas can be expressed in the form $y = mx + b$. Since a general law has many specific applications, the linear function, being a general law, can be applied to the study of numerous relations. In this chapter we shall study the properties and applications of the function $y = mx + b$.

Graph of the Linear Function

Linear is an appropriate name for the function $y = mx + b$, since the graph of this function is a straight line. Although "line" is one of the basic undefined terms of Euclidean geometry, everyone intuitively feels he knows what is meant by "straight line." This notion is intimately related to the idea of direction of a line. One way of specifying direction is given in the following definition.

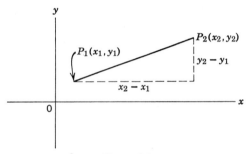

Figure 6.1

Definition 1. Given two points $P_1(x_1, y_1)$ and $P_2(x_2, y_2)$, the slope of the line segment connecting the two points is

$$\frac{y_2 - y_1}{x_2 - x_1}$$

The graphic interpretation of the slope is illustrated in Figure 6.1. The slopes of specific line segments, illustrated in Figure 6.2, are computed as follows:

$$\text{slope of } AC = \frac{4-1}{5-2} = \frac{3}{3} = 1$$

$$\text{slope of } AB = \frac{5-1}{10-2} = \frac{4}{8} = \frac{1}{2}$$

$$\text{slope of } AE = \frac{1-1}{12-2} = \frac{0}{10} = 0$$

$$\text{slope of } AD = \frac{-3-1}{4-2} = \frac{-4}{2} = -2$$

Thus we observe that the slope of a horizontal line is zero, and that as the slope increases through positive values, the line appears to rotate counterclockwise toward a limiting vertical position. Similarly, as the slope decreases through negative values the line appears to rotate clockwise toward a limiting vertical position. The slope, therefore, becomes a numerical index associated with specific direction of a given line. The slope of a vertical line is undefined.

Another useful interpretation of the numerical value of the slope is that it represents the number of units y changes per unit increase in x. Thus, a slope of -2 means that y decreases 2 units per unit increase in x. It is left to the reader to verify that this interpretation is a direct consequence of the geometric properties of similar triangles.

To support the statement that the graph of $y = mx + b$ is a straight line, we now demonstrate that the slope of the line segment joining *any* two points

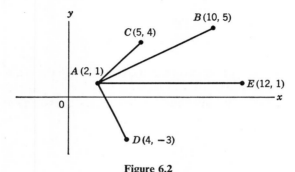

Figure 6.2

on the graph of the function is always the same. The procedure is fundamentally algebraic. If we select $P_1(x_1, y_1)$ and $P_2(x_2, y_2)$ to be any two points on the graph, then each of these ordered pairs of coordinates must satisfy the given equation. Therefore

$$y_2 = mx_2 + b \quad \text{and} \quad y_1 = mx_1 + b$$

Then

$$\frac{y_2 - y_1}{x_2 - x_1} = \frac{(mx_2 + b) - (mx_1 + b)}{x_2 - x_1} = \frac{m(x_2 - x_1)}{x_2 - x_1} = m$$

We thus not only prove that the slope of the line joining any two points of the graph is constant, but also that this constant is the value m in the equation $y = mx + b$.

With this discovery it is natural to inquire whether the remaining constant, b, also has any direct relation to the graph of the function. We note that when $x = 0$, $y = b$. Therefore, b represents the y-intercept of the function, that is, the graph intersects the y-axis at the point $(0, b)$. In summary, then, the graph of the function $y = mx + b$ is a straight line whose slope is m and whose y-intercept is b.

What is a Linear Function?

EXAMPLE 1. Discuss the graph of the function $y = -5x + 2$.

Solution: The graph is a straight line, since the equation is of the form $y = mx + b$. The line intersects the y-axis at the point $(0, 2)$ and has a slope of -5. This means that y decreases 5 units as x increases 1 unit anywhere along the line.

EXAMPLE 2. If the graph of $y = f(x)$ is a nonvertical straight line, prove that $y = mx + b$.

Solution: This is the converse of the statement that the graph of $y = mx + b$ is a straight line. Since the line is nonvertical, we can assume that it intersects the y-axis at a point $(0, b)$ and has a slope m. Let $P(x, y)$ be any point on the line. (See Figure 6.3.) Then, by the definition of slope,

$$m = \frac{y - b}{x - 0}$$
$$mx = y - b$$
$$y = mx + b$$

EXAMPLE 3. What is the equation of the line through the point $(4, 1)$ whose slope is 2?

Solution: Let $P(x, y)$ be any point on the line. Using the definition of slope, we have,

$$2 = \frac{y - 1}{x - 4}$$
$$2x - 8 = y - 1$$
$$y = 2x - 7$$

EXAMPLE 4. The two points $P_1(3, 2)$ and $P_2(5, -4)$ determine a line. What is the equation of the line?

Solution: The slope of the line segment connecting the two points is

$$m = \frac{-4 - 2}{5 - 3} = \frac{-6}{2} = -3$$

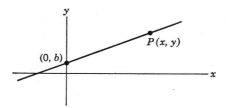

Figure 6.3

Let $P(x, y)$ be any point on the line segment or its extension. Then the slope between the points P and P_1 must also equal -3. Therefore,

$$-3 = \frac{y - 2}{x - 3}$$
$$-3x + 9 = y - 2$$
$$y = -3x + 11$$

The required equation is $y = -3x + 11$. Note that both the given points satisfy this equation.

EXERCISE 32

The Linear Function

Without graphing, determine the slope and y-intercept of the line represented by each of the following functions.

1. $y = 2x + 1$
2. $y = 1/2\, x - 5$
3. $y = x$
4. $y = 6$
5. $y = -1/4\, x + 9$
6. $y = x - 1$

Graph each of the following functions by locating the y-intercept and then determining a second point with the use of the slope. Verify your result by checking whether a third point on the line satisfies the equation.

7. $y = x - 2$
8. $y = 1/2\, x + 3$
9. $y = 2x - 1$
10. $y = -2x + 5$
11. $y = -x$
12. $y = 3$

Graph the straight line that satisfies the following conditions and determine its equation (Problems 13–20).

13. Crosses the y-axis at $(0, 5)$ with a slope of -2.
14. Passes through the point $(4, 3)$ with a slope of 3.
15. Intersects the y-axis at $y = 3$ with a slope of $1/2$.
16. Passes through the points $(1, 5)$ and $(4, 14)$.
17. Passes horizontally through the point $(3, -2)$.
18. Passes through the origin with a slope of 2.
19. Contains the points $(0, 3)$ and $(5, 0)$.
20. Intersects the y-axis at the origin and passes through the point $(-2, 8)$.
21. For a fixed value of b, the equation $y = mx + b$ determines a family or system of lines passing through the point $(0, b)$. On a single graph, illustrate the family of lines through the point $(0, 3)$ for $m = -8, -6, -4, -2, 0, 2, 4, 6, 8$.
22. For a fixed value of m, the equation $y = mx + b$ determines a system of parallel lines each having a slope m. On a single graph, illustrate the family of lines whose slope is $1/2$ for $b = -2, -1, 0, 1, 2, 3, 4$.

Determine the equation of the linear function that satisfies each of the following conditions. (Problems 23–25.)

23.

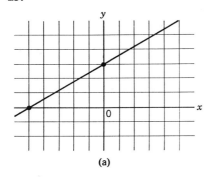

(a)

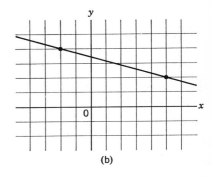

(b)

24.

x	-2	-1	0	1	2	3
y	3	5	7	9	11	13

25.

x	-4	-2	0	2	4	6
y	8	7	6	5	4	3

33. RATE OF CHANGE OF LINEAR FUNCTIONS

The average rate of change of a function has been defined by the expression

$$\frac{f(x+h) - f(x)}{h}$$

If we apply this expression to the linear function $y = mx + b$, we obtain:

$$\text{average rate of change} \Big|_{x}^{x+h} = \frac{m(x+h) + b - (mx+b)}{h}$$

$$= \frac{mx + mh + b - mx - b}{h} = \frac{mh}{h} = m$$

Thus, for a linear function the average rate of change is constant and equal to the slope m. Since the average rate of change is constant in any interval h, the instantaneous rate of change is also constant. Therefore, we need not differentiate between the rate of change at any instant and the average rate of change over an interval in the case of a linear function. We will merely refer to the rate of change of the function $y = mx + b$.

The fact that the rate of change is equal to the slope is not just a fortuitous accident—after all, the very definition of slope made this predictable. It will

33. Rate of Change of Linear Functions

be recalled that, numerically, the slope indicates the number of units y changes per unit increase in x. Similarly, the rate of change indicates the change in the dependent variable per unit increase in the independent variable. Thus, the slope is a graphic interpretation of the rate of change of a linear function.

We now illustrate the application of these ideas to a specific function. Consider the formula $F = 9/5\ C + 32$, which relates temperature readings on a Fahrenheit and centigrade scale. This is a linear function, and its graph is shown in Figure 6.4. We note from the formula that the slope is 9/5. One verification of this is the calculation of slope between the freezing point of water $(0, 32)$ and its boiling point $(100, 212)$. The slope between these two points is

$$\frac{212 - 32}{100 - 0} = \frac{180}{100} = \frac{9}{5}$$

Similarly, the slope between any other two readings is 9/5. The physical interpretation, of course, is that the Fahrenheit reading will increase $1\frac{4}{5}°$ for each degree increase in centigrade reading.

Two additional graphic characteristics of a function have important physical interpretations. The y-intercept is frequently referred to as the *initial value of the function* because it represents the value of the function when the independent variable is zero. The x-intercept is called the *zero of the function* because it is the value of the independent variable which makes the value of the function zero. We observe that the initial value of the function

$$F = \tfrac{9}{5}C + 32$$

is 32 (determined by setting $C = 0$ and solving for F), and the function has a zero at $-17\frac{7}{9}$ (determined by setting $F = 0$ and solving for C).

We have thus established the following relations for any function which can be expressed in the form $y = mx + b$.

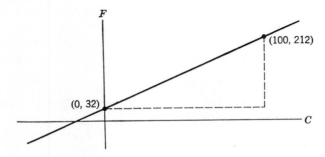

Figure 6.4

What is a Linear Function?

1. The rate of change of the function = slope = m.
2. The initial value of the function = y-intercept = b.
3. The zero of the function = x-intercept.

EXAMPLE 1. If $1 is invested at 5% simple interest, its total value at the end of t years is given by $A = 0.05t + 1$. Determine the rate of change of the function represented by this equation, the initial value of the function, and the zero of the function.

Solution: Since the relation $A = 0.05t + 1$ is of the form $y = mx + b$, it represents a linear function. Therefore its rate of change is $m = 0.05$, which means that the amount A increases 0.05 dollars when the time increases 1 year. The initial value of the function is $b = 1$, which is the amount at $t = 0$. The zero of the function can be found by setting $A = 0$ and solving for t:

$$0 = 0.05t + 1$$
$$t = -\frac{1}{0.05} = -20$$

Since for practical purposes the domain of the function is limited to the set of positive real numbers, this value of t must be discarded. Therefore, the function has no zero in the domain of positive real numbers.

EXERCISE 33

RATE OF CHANGE OF LINEAR FUNCTIONS

Answer each of the following questions for Problems 1–10.

(a) What is the rate of change of the function and what does it represent?
(b) What is the initial value of the function and what does it represent?
(c) What is the zero (if any) of the function and what does it represent?
(d) What is the value of $f(4)$ and what does it represent?

1. A freight car is moving in such a way that its distance from a given loading point is given by the relation $s = 45t + 360$, where t is measured in hours and s is measured in miles.

2. When a variable force of w pounds is applied to an elastic cord fastened at one end, the resulting length of the cord is expressed by $s = 1/2\ w + 6$, where s is measured in inches.

3. The amount of fuel remaining in the tanks of an air-cargo plane is given by $G = -70t + 400$, where G is the amount of fuel in gallons and t is the flying time in hours.

4. The length of a steel rod is found to vary with its temperature in such a way that $L = 0.0007C + 95.22$, where L is measured in feet and C is the centigrade reading of temperature.

5. The velocity of a ball rolled up an inclined plane is given by $v = -6t + 30$, where v is measured in feet per second and t is measured in seconds.
6. For relatively low altitudes above sea level the atmospheric pressure is approximated by the relation $p = -0.06h + 15$, where p is the pressure in pounds per square inch and h is the altitude in hundreds of feet.
7. When an electrical current is passed through a 5 ohm resistor, the relation between the voltage E and the amperage I is given by $E = 5I$.
8. An appliance shop bases its charges for service calls on the formula $C = 7.50t + 5$, where C is the charge in dollars and t is the time in hours required by the serviceman to complete his repairs.
9. On a Reaumer thermometer $R = 0$ marks the freezing point and $R = 80$ marks the boiling point of water. The relation of Reaumer readings to Fahrenheit readings is given by the formula $F = 9/4\, R + 32$.
10. In a certain locality the total cost of site and building for a warehouse is approximated by the formula $C = 2000x + 30{,}000$ where x is the gross area of the warehouse in hundreds of square feet and C is the cost in dollars.

Figure 6.5 shows the velocity of a car t seconds after its brakes have been applied. Problems 11–15 refer to this graph.

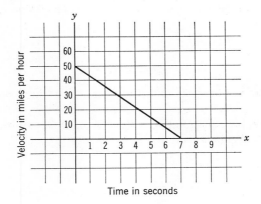

Figure 6.5

11. What is the rate of change of the function represented by the graph? What does it represent?
12. What is the initial value and what does it represent?
13. What is the zero of the function and what does it represent?
14. Find $f(3)$ from the graph. What does it represent?
15. Using the slope and the y-intercept, write the equation of the function.

In studying the elasticity of a new synthetic fiber, a laboratory team tabulated the following results of an experiment:

x = tension in pounds	0	2	4	6	8	10
y = length of strand in inches	24.00	24.18	24.36	24.54	24.72	24.90

16. What is the rate of change of the function?
17. What is the initial value of the function and what does it represent?
18. Does this function have a zero? Explain.
19. Express y as a function of x.

34. SYSTEMS OF LINEAR FUNCTIONS

The Linear Form $Ax + By = C$

The linear function often appears in the form $Ax + By = C$. For example, $6x + 2y = 1$ is a linear function. To rewrite this equation in the familiar form $y = mx + b$, we can solve the given equation for y as follows:

$$6x + 2y = 1$$
$$2y = -6x + 1$$
$$y = -3x + \tfrac{1}{2}$$

Therefore, the graph of $6x + 2y = 1$ is a straight line whose slope is -3 and whose y-intercept is $1/2$.

In the same way, the equation $Ax + By = C$, $B \neq 0$, can be written in the form $y = mx + b$. Again solving for y, we have

$$Ax + By = C$$
$$By = -Ax + C$$
$$y = -\frac{A}{B}x + \frac{C}{B}$$

which is of the form $y = mx + b$, where the slope $m = -(A/B)$ and the y-intercept is C/B.

Since a straight line is determined by any two points, the graph of $Ax + By = C$ can usually be drawn conveniently by locating its x and y intercepts. A third point, if desired, can be plotted to ensure against errors and to promote graphic accuracy by obtaining points which are spaced sufficiently far apart.

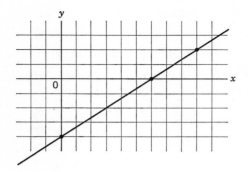

Figure 6.6

EXAMPLE 1. Graph the function defined by $2x - 3y = 12$.
  Solution: For $x = 0$, $-3y = 12$ from which $y = -4$. For $y = 0$, $2x = 12$ from which $x = 6$. The results are shown in the following table. The third point is plotted as a partial check of the intercepts (see Figure 6.6.)

Systems of Two Linear Functions

We now complicate the situation slightly by considering a pair of linear functions, that is, a *system* of two equations of the form

$$A_1 x + B_1 y = C_1$$
$$A_2 x + B_2 y = C_2$$

The *solution set of the system* is defined to be the set of all ordered pairs (x, y) that satisfy both equations. Graphically, the solution set consists of all points that lie on both of the two given lines. This graphic interpretation has strong implications concerning the nature of the solution set. If the two lines are plotted in the coordinate plane, one of three possible situations must exist.

1. They intersect, in which case they have a single point in common. The solution set consists of a single ordered pair of numbers.

2. They are parallel, in which case they have no points in common. The solution set is a null or empty set.

3. They coincide, that is, they are really one and the same straight line. Hence, every point on the common line is a solution so that the solution set is infinite.

Thus the solution set may consist of a single solution, no solution, or infinitely many solutions. The following examples illustrate how these cases arise algebraically.

EXAMPLE 2. Find the solution set of the system

$$x - 3y = 1$$
$$3x + 2y = 14$$

Solution: We first eliminate either x or y by the procedure of adding equalities. To eliminate x, we multiply the first equation by 3 and the second by -1 to obtain the equivalent system

$$3x - 9y = 3$$
$$-3x - 2y = -14$$

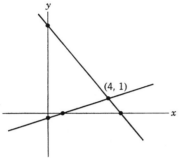

Figure 6.7

Adding, we have $-11y = -11$, or $y = 1$. Since the solution set consists of all ordered pairs belonging to either line, which make the equality $y = 1$ a true statement, we may substitute $y = 1$ in either equation and solve for x. We find that $x = 4$ and, therefore, the solution set consists of the ordered pair (4, 1). (See Figure 6.7.)

EXAMPLE 3. Examine the solution set of the system

$$2x + 5y = 15$$
$$4x + 10y = -8$$

Solution: To eliminate x, as in Example 2, we form the equivalent system

$$4x + 10y = 30$$
$$-4x - 10y = 8$$

Adding these equalities, we obtain the impossible result $0 = 38$. Since there is no ordered pair belonging to either line which makes the relation $0 = 38$ a true statement, there is no solution. The lines are parallel, as illustrated in Figure 6.8.

EXAMPLE 4. Solve the system

$$-3x + y = -7$$
$$9x - 3y = 21$$

Solution: Multiplying the first equation by 3 and rewriting the second equation, we have the equivalent system

$$-9x + 3y = -21$$
$$9x - 3y = 21$$

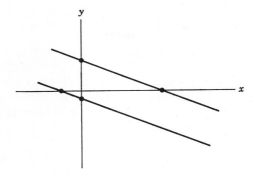

Figure 6.8

Adding these equalities, we obtain $0 = 0$. Therefore, the equivalent system consists of two expressions for the same equation. In effect, there is a single equation, that is, the lines are coincident and there are infinitely many solutions. (See Figure 6.9.) We can list as many of these as we please by merely selecting values of x and computing the corresponding values of y. Thus, for $x = 1, 2, 3, 4, 5$, we have the ordered pairs $(1, -4)$, $(2, -1)$, $(3, 2)$, $(4, 5)$, $(5, 8)$, respectively.

Returning now to the general case

$$A_1 x + B_1 y = C_1$$
$$A_2 x + B_2 y = C_2$$

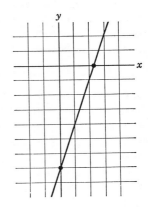

Figure 6.9

we note that the slope of the first line is $-(A_1/B_1)$, and the slope of the second line is $-(A_2/B_2)$. If there is a single solution the two slopes will be unequal, that is, the two lines will intersect. This will happen whenever $A_1/B_2 \neq A_2/B_2$ or, equivalently, $A_1/A_2 \neq B_1/B_2$. (See Example 2.)

If $A_1/A_2 = B_1/B_2 \neq C_1/C_2$, then the slopes will be equal but the y-intercepts will be unequal. Therefore, the system consists of two parallel, distinct lines which have no points in common. The system has no common solution, as in Example 3.

If $A_1/A_2 = B_1/B_2 = C_1/C_2$, the slopes and intercepts are equal. Consequently the lines are coincident and the system has an unlimited number of solutions, as illustrated in Example 4.

Family of Lines Through the Intersection of Two Lines

The summary of the general case, together with the graphic interpretations and algebraic procedures, is a good example of how algebra and geometry reinforce each other in the study of functions. This interplay is particularly significant if we carry it one step further to consider what actually takes place graphically when we multiply one or both of the given equations by a constant and add the resulting equalities.

Let the equations

$$a_1 x + b_1 y = c_1$$
$$a_2 x + b_2 y = c_2 \tag{1}$$

represent two lines which intersect at the point $P(x_0, y_0)$. Then the replacements (x_0, y_0) must satisfy both equations, that is,

$$a_1 x_0 + b_1 y_0 = c_1$$
$$a_2 x_0 + b_2 y_0 = c_2 \tag{2}$$

When we eliminate one unknown, we usually multiply one of the equations by some integral constant k_1 and the other equation by some integer k_2, and then add the resulting equations. There is no loss of generality if we merely multiply one equation by a rational number k and add it to the remaining equation to obtain

$$(a_1 x + b_1 y) + k(a_2 x + b_2 y) = c_1 + k c_2 \tag{3}$$

Now this equation always represents a straight line—in particular, a line which, for every value of k, passes through the point of intersection $P(x_0, y_0)$. To verify this, note that substituting x_0 for x and y_0 for y in Equation 3 will satisfy the equation, since the first parenthetical expression will equal c_1 and

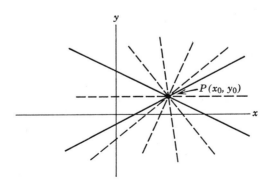

Figure 6.10

the second will equal c_2. In effect, Equation 3 represents a *family* of lines, each of which passes through the point $P(x_0, y_0)$, as illustrated in Figure 6.10. When we eliminate x, we are actually finding the one member of this family which is horizontal, that is, the line $y = y_0$. Similarly, when we eliminate y from the given equations, we are finding the one member of this family that is vertical. This produces the solution $x = x_0$, $y = y_0$.

Although this development is mathematically significant and intellectually challenging, it has further merit in being useful in finding solutions to a number of other problems. The following example illustrates one such application.

EXAMPLE 5. What is the equation of the line that passes through the intersection of

$$x + 3y = 7$$
$$2x - y = 5 \tag{1}$$

and has a slope of -5?

Solution: It is possible, of course, to determine the point of intersection and then use the methods of Section 32 to determine the equation of the line through that point which has the given slope. The solution is somewhat more elegant if we avoid finding the point of intersection but instead write the equation of the family of lines through the point of intersection:

$$(x + 3y) + k(2x - y) = 7 + 5k \tag{2}$$

We rewrite this in the form $Ax + By = C$:

$$(1 + 2k)x + (3 - k)y = 7 + 5k \tag{3}$$

The slope of a line expressed in this form is $-(A/B)$, which in this case is $-\left(\dfrac{1 + 2k}{3 - k}\right)$. Therefore

$$-\frac{1 + 2k}{3 - k} = -5$$
$$1 + 2k = 15 - 5k$$
$$7k = 14$$
$$k = 2$$

Hence, for $k = 2$, a member of the given family of lines will have the required slope. Substituting this value of k in Equation 2, we have

$$x + 3y + 2(2x - y) = 7 + 10$$
$$5x + y = 17$$

which is the required equation. This procedure is particularly useful in its application to higher degree functions.

EXERCISE 34

Systems of Linear Functions

Without solving, determine the number of common solutions which exist for each of the following systems of equations.

1. $x + 2y = 3$
   $2x + 2y = 6$
2. $x + y = 7$
   $x - y = 7$
3. $3x - y = 5$
   $6x - 2y = 5$
4. $x - 7y = 1$
   $2x - 14y = 7$
5. $3x - 3y = 6$
   $4x - 4y = 8$
6. $2x - y = 1$
   $x - 2y = 1$
7. $x - 2y = 3$
   $-x + 2y = 3$
8. $3x - 2y = -5$
   $-9x + 6y = 15$
9. $2x - 3y = 9$
   $3x - 2y = 9$

Solve the following equations algebraically. Then plot the graph and verify your solution graphically.

10. $x + y = 0$
    $2x - 3y = 15$
11. $3x + y = 2$
    $2x - y = 1$
12. $2x - 2y = 7$
    $x - y = 7$
13. $4x - y = 15$
    $3x - y = 9$
14. $8x - 6y = 12$
    $4x - 3y = 6$
15. $x - y = 6$
    $x + y = 8$
16. $4x - 2y = 6$
    $2x - y = 3$
17. $x - y = -3$
    $x + y = 11$
18. $2x + y = 6$
    $2x + y = 0$
19. $x + 7y = 21$
    $-x + y = 11$
20. $x + 2y = 11$
    $2x + y = 10$
21. $5x + 2y = 10$
    $15x + 6y = 60$

Without finding the point of intersection of the two given lines, write the expression for the equation of all lines through the point common to:

22. $x + 2y = 5$
    $2x - y = 3$
23. $x + y = 1$
    $x - y = 0$
24. $2x + y = 1$
    $x + 3y = 1$
25. $x + 2y = 0$
    $y = 5$
26. $3x - y = 2$
    $x - y = 1$
27. $x - y = 9$
    $y = 8$

Without finding the point of intersection of the two given lines, write the equation of the one line passing through the point common to

$$11x + 2y = 1$$
$$x + y = 1$$

and having a slope equal to:

28. -4
29. -7
30. $1/2$

35. APPLICATIONS OF LINEAR SYSTEMS

When a verbal problem concerns two unknown quantities, it is often expedient to use a separate variable to represent each unknown. The conditions of the problem must then be interpreted to provide two equations involving the variables. If the resulting system of equations has a common solution which checks against the stated requirements, the problem is solved; however, if the system has no common solution or if the common solution fails to satisfy the conditions of the problem, then no solution is possible. The following examples illustrate the procedure and the interpretation of results.

EXAMPLE 1. A manufacturer of electronic components finds that his cost of fabricating a certain item is $C = kn + f$, where n is the number of units to be produced, k is his operating cost per unit, and f is his fixed cost for tooling up, making prototypes, etc. The cost for 12,000 units is \$13,200, and the cost for 20,000 units is \$19,200. Find the fixed cost and the operating cost per unit.

Solution: Substituting the given values in the formula results in the system

$$13{,}200 = 12{,}000k + f$$
$$19{,}200 = 20{,}000k + f$$

Multiplying the first equation by -1 and adding the result to the second equation, we have

$$6000 = 8000k$$
$$k = 0.75$$

Substituting this value of k in either equation of the given system, we find $f = 4200$. Therefore, the fixed cost is \$4200 and the operating cost per unit is 75 cents.

EXAMPLE 2. Two elastic cords are 18 inches and 23 inches long, respectively. The first stretches 0.3 inch for each pound of force applied, and the second stretches 0.5 inch for each pound of force. Can equal forces be applied to each cord so that they will have the same length?

Solution: Let x = number of pounds of force to be applied to each cord
y = final length of the cords, in inches

The equations are

$$y = 18 + 0.3x \text{ (final length of first cord}$$
$$\text{equals its initial length plus its stretch)}$$

$$y = 23 + 0.5x \text{ (final length of second cord}$$
$$\text{equals its initial length plus its stretch)}$$

Solving these equations, we find

$$x = -25$$

The system has a single solution, but it must be rejected because the problem does not have meaning for negative values of force. Therefore, the problem has no solution, that is, equal forces cannot be applied to each cord to equalize their lengths.

EXAMPLE 3. The sum of the digits of a two-digit hexal (base 6) number is 5. If the digits are reversed, the new number is 5 less than the original. Find the number.
Solution: Let x = digit in the six's place
y = digit in the unit's place
Then, in terms of ordinary decimal notation,

$$6x + y = \text{value of the original number}$$

$$6y + x = \text{value of the number obtained by reversing the digits}$$

The equations are

$$x + y = 5 \text{ (the sum of the digits is 5)}$$
$$(6x + y) - (6y + x) = 5 \text{ (the new number is 5 less than the original)}$$

Rewriting the second equation, the system becomes

$$x + y = 5$$
$$5x - 5y = 5$$

Solving this system leads to the values

$$x = 3, y = 2$$

Therefore, the number is 32_6. This number satisfies the conditions of the problem, since the sum of the digits is 5 and, reversing the digits, 23_6 is 5 less than 32_6.

EXERCISE 35

APPLICATIONS OF LINEAR SYSTEMS

Solve each of the following problems by introducing two variables.

1. The total capacity of two storage tanks is 1985 gallons. If one tank holds 515 gallons more than the other, find the capacity of each tank.
2. In the equation $y = mx + b$, it is known that $y = 20$ when $x = 3$, and $y = -8$ when $x = -1$. Find the values of m and b.
3. The two acute angles of a right triangle total 90°. If the larger of these angles is 15° more than twice the smaller, find the size of each acute angle.
4. A rectangular piece of metal has a perimeter of 18 inches. Find its dimensions if the length exceeds the width by 6 inches.
5. A decimal number between 10 and 100 is increased by 45 if its digits are reversed. Find the number if the sum of its digits is 11.
6. The sum of the two digits of a base 5 number is 11_5. If 13_5 is subtracted from the number, the digits will be reversed. Find the number.
7. The total cost of two bonds is $3500. One of the issues, at 6% return, yields $89 more per year than the other which has a 5% return. Determine the cost of each bond.
8. Separate 119 into two parts such that the larger is six times the smaller.
9. Natural gas for space heating is supplied at a fixed monthly service charge plus a given rate per thousand cubic feet of gas. A family is billed $15.90 for the month of November when they used 160,000 cubic feet, and $29.40 for the following month when they used 310,000 cubic feet. Find the amount of the service charge and the rate per thousand cubic feet.
10. A tourist exchanges 20 coins, consisting of shillings and marks, for $4.01. If the exchange rate for shillings is 14 cents and for marks is 25 cents, find the number of coins of each kind involved in the exchange.
11. A reserve tank has a capacity of 600 gallons. Fuel is pumped into it through two pipes of which the larger delivers 4 gallons more per minute than the smaller. If the tank is filled by allowing both pipes to deliver for 20 minutes, find the rate of delivery of each pipe.
12. A consultant and his assistant receive a fee of $365 for a project which required 2 days of the consultant's time and 3 days of his assistant's time. Another project involved the consultant 3 days and his assistant 4 days for a fee of $520. What was the daily fee for each man?
13. The weekly market demand for a certain commodity is given by

$$x = 15 - 2p$$

where p is the price index of the commodity and the weekly market supply is

given by
$$x = 3p - 3$$

For what value of x will the demand price equal the supply price? (In the study of economics this is known as the equilibrium market price.)

14. A manufacturer has a work force of 825 men and 375 women. Product A requires twice as many men as women for its production, while product B requires three times as many men as women. How should the work force be apportioned between the two products if it is to be fully utilized?

15. A chemist has a supply of 90% solution of nitric acid and 70% solution of the same acid. He receives a request for 100 gallons of a 75% solution. How many gallons of each should he mix to fill the request?

7

What is a Quadratic Function?

36. THE QUADRATIC FUNCTION

If a ball is thrown with an upward velocity of 100 feet per second from the top of a building 30 feet high, the approximate height, s, of the ball above the ground t seconds after it is thrown is given by the formula

$$s = -16t^2 + 100t + 30 \tag{1}$$

Equation 1 is an example of an important class of functions, called *quadratic functions*, which we now define.

Definition 1. A quadratic function is a set of ordered pairs (x, y) where $\{(x, y) \mid y = ax^2 + bx + c\}$. The domain of x is the set of real numbers; a, b, and c are constants, and $a \neq 0$. The condition $a \neq 0$ is stated to avoid the possibility of the function being linear and not quadratic. Examples of quadratic functions are:

(a) $y = x^2 - 3x + 5$
(b) $y = 3x^2 - 2x$
(c) $y = 2x^2 - 6$
(d) $y = -16x^2 + x$
(e) $y = 4x^2$

Graph of the Quadratic Function

Several properties of the quadratic function may be illustrated by its graph. As an example, let us construct the graph of the quadratic function

$$y = 2x^2 - 4x - 6$$

It is convenient to make a table of ordered pairs as follows:

x	-2	-1	0	1	2	3	4
y	10	0	-6	-8	-6	0	10

The points represented by the ordered pairs are joined by a smooth curve, shown in Figure 7.1. We make several observations from the graph.

1. The curve, called a *parabola*, has a lowest, or *minimum*, value of y a the point $(1, -8)$. This minimum point is called the *vertex* of the parabola

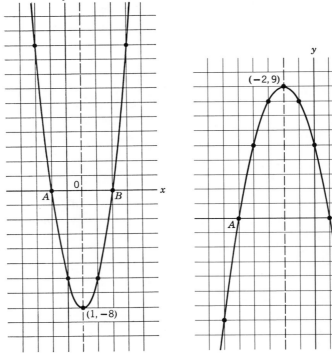

Figure 7.1   Figure 7.2

2. The parabola is symmetric with respect to a line parallel to the y-axis and passing through the vertex. This line, whose equation is of the form $x = k$, is called the *axis of symmetry*. The particular value of k is the x-coordinate of the vertex. The axis of symmetry of this parabola is $x = 1$.

3. The parabola, for this particular function, opens upward.

4. The curve has two x-intercepts at $A(-1, 0)$ and $B(3, 0)$. As in Section 33, the x-intercepts of a function are also referred to as the *zeros of the function*. Thus this function has two distinct zeros, at $x = -1$ and $x = 3$.

As a second example, we construct the graph of the quadratic function

$$y = -x^2 - 4x + 5$$

A suitable table of values is made as follows:

x	-6	-5	-4	-3	-2	-1	0	1	2
y	-7	0	5	8	9	8	5	0	-7

The graph is shown in Figure 7.2. We observe from the graph that for this function the parabola has a highest, or *maximum*, value of y at the point $(-2, 9)$, since the curve opens downward. The axis of symmetry is the line $x = -2$. The curve has two real zeros, A and B, which are the points $(-5, 0)$ and $(1, 0)$.

It will be shown later that for the quadratic function $y = ax^2 + bx + c$:

1. If $a > 0$, there is a minimum point and the curve opens upward.

2. If $a < 0$, there is a maximum point, and the curve opens downward.

It is important to note that a quadratic function may have no real zeros. In Figure 7.3, which is the graph of $y = 2x^2 - 4x + 8$, it is clear that since the minimum point is $(1, 6)$ and the curve opens upward, it will not cross the x-axis and, hence, will have no real zeros.

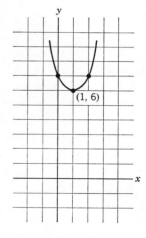

Figure 7.3

The Maximum or Minimum Value of a Quadratic Function

We shall now prove that the quadratic function $y = ax^2 + bx + c$ has a maximum or a minimum value at the point whose x-coordinate is $-(b/2a)$. To do this, we shall complicate the form of the function, but the results will justify this procedure. We assume that the reader will follow the algebraic process of each step.

$$y = ax^2 + bx + c \tag{1}$$

Factoring out a,

$$y = a\left(x^2 + \frac{b}{a}x + \frac{c}{a}\right) \tag{2}$$

Adding and subtracting $b^2/4ac$ to make a perfect square,

$$y = a\left(x^2 + \frac{b}{a}x + \frac{b^2}{4a^2} - \frac{b^2}{4a^2} + \frac{c}{a}\right) \tag{3}$$

Putting the last three terms over a common denominator,

$$y = a\left(x^2 + \frac{b}{a}x + \frac{b^2}{4a^2} - \frac{b^2 - 4ac}{4a^2}\right) \tag{4}$$

Grouping terms as the difference of two squares,

$$y = a\left[\left(x + \frac{b}{2a}\right)^2 - \left(\frac{\sqrt{b^2 - 4ac}}{2a}\right)^2\right] \tag{5}$$

We observe from Equation 5 that since the quantity $[x + (b/2a)]^2$ is either zero when $x = -(b/2a)$, or positive when $x \neq -(b/2a)$, the quantity in the brackets has the least value when $x = -(b/2a)$. Hence, when $a > 0$, y assumes its least value when $x = -(b/2a)$, and when $a < 0$, y assumes its greatest value when $x = -(b/2a)$. In either case, when $x = -(b/2a)$, $y = -[(b^2 - 4ac)/4a]$, which is either a maximum or minimum value of the function. Hence the coordinates of the maximum or minimum point of the parabola, that is, the vertex, are

$$\left(-\frac{b}{2a}, -\frac{b^2 - 4ac}{4a}\right) \tag{6}$$

In practice, it is usually easier to determine the y coordinate of the maximum or minimum point by first finding the x-value, $-(b/2a)$, and then substituting this value of x in the function. One more observation is worthy of note. Since a value of a greater than zero indicates that the curve has a minimum point, it follows that the parabola opens upward for $a > 0$. In a similar way we may show that the parabola opens downward when $a < 0$.

EXAMPLE 1. Find the coordinates of the maximum point of the curve $y = -2x^2 - 6x + 7$, and give the equation of the axis of symmetry.

Solution: The maximum point of the curve has the x-coordinate, $-b/2a$, or $6/-4 = -(3/2)$. For $x = -(3/2)$, $y = 23/2$. Hence the coordinates of the vertex are $[-(3/2), 23/2]$, and the equation of the axis of symmetry is $x = -(3/2)$.

EXAMPLE 2. Find the coordinates of the minimum point of the curve $y = 2x^2 - 8x + 5$, and give the equation of the axis of symmetry.
Solution: The minimum point of the curve has the x-coordinate $8/4 = 2$. For $x = 2$, $y = -3$. Hence the coordinates of the vertex are $(2, -3)$ and the equation of the axis of symmetry is $x = 2$.

EXAMPLE 3. Show that the graph of $y = x^2 - x + 1$ has no real zeros.
Solution: Since $a > 0$, the curve opens upward. The minimum point of the curve is $(1/4, 3/4)$, which is above the y axis. Hence the curve cannot cross the x-axis, and the function has no real zeros.

Sketching the Quadratic Function

The foregoing discussion furnishes us with a rapid means of sketching the graph of the quadratic function.

1. Find and plot the y-intercept.
2. Find and plot the coordinates of the maximum or minimum point.
3. Draw the axis of symmetry of the curve.
4. Find and plot a point symmetrical to the y-intercept with respect to the axis of symmetry.
5. The three points thus determined are sufficient to sketch the curve. Additional points may be plotted, particularly in the neighborhood of the vertex and the zeros, if necessary.

EXAMPLE 4. Sketch the graph of the quadratic function $y = 2x^2 + 3x - 5$.
    Solution: 1. The y-intercept is $(0, -5)$.
            2. The minimum point is $[-(3/4), -49/8]$.
            3. The axis of symmetry is the line $x = -(3/4)$.

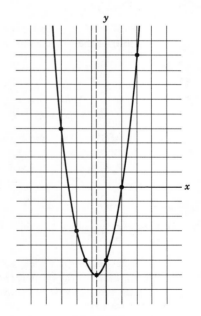

Figure 7.4

4. The point symmetric to $(0, -5)$ is $[-(3/2), -5]$.
5. Although we have enough information to sketch the curve, the following additional points will help us to refine the graph.

x	-3	-2	-1	0	1	2
y	4	-3	-6	-5	0	9

The graph of the function is illustrated in Figure 7.4.

EXERCISE 36

THE QUADRATIC FUNCTION

Which of the following functions are quadratic functions? Letters other than y, x, or t designate constants.

1. $y = 2x^2 - 7x - 3$
2. $y = 9x - 7$
3. $y = ax^2 + bx + k$
4. $y = x^2$
5. $y = 6t^2 - 7$
6. $y = (6 - k)x^2 - kx + m$
7. $y = -3x^2 + 2x - 3$
8. $y = (m - n)t^2$
9. $y = (a^2 - b^2)x + c$
10. $y = \sqrt{3}\, t^2 + \sqrt{2}\, t + 6$

State the value of a, b, and c for each of the following quadratic functions. (If necessary, rewrite the function in the form $y = ax^2 + bx + c$.)

11. $y = 3x^2 - 7x + 2$
12. $y = x^2 - 17x - 7$
13. $y = (m - n)x^2 + (m + n)x + 3$
14. $y = 6x^2 - mx^2 + 7x - nx + 3$ (*Hint.* Rewrite as $(6 - m)x^2$ etc.)
15. $y = 3x^2 - kx^2 + 7x + k + 3$
16. $y = 2mx^2 - 3x + mx + m$

For each of the following parabolas, (a) determine the coordinates of the vertex, (b) state whether the vertex is a maximum or minimum point of the curve, and (c) write the equation of the axis of symmetry.

17. $y = 2x^2 - 8x + 3$
18. $y = x^2 - 3x + 4$
19. $y = x^2 + 3x - 4$
20. $y = 3x^2 - 7x + 2$
21. $y = -2x^2 + 3x - 2$
22. $y = -6x^2 - 24x - 13$

Sketch the graph of each of the following quadratic functions. For each function, give the:

(a) Coordinates of the vertex.
(b) Coordinates of the y-intercept.
(c) Equation of the axis of symmetry.
(d) Coordinates of the point symmetric to the y-intercept.
(e) Zeros (estimated) if they exist.

23. $y = x^2 - 2x + 1$ 24. $y = x^2 - 8x + 4$ 25. $y = x^2 + 4x - 5$
26. $y = 2x^2 - 7x$ 27. $y = -x^2 + 10x + 4$ 28. $y = -2x^2 - 12x + 9$
29. $y = x^2 - 9$ 30. $y = 3x^2 - 14$ 31. $y = 1/2\, x^2 - 3x + 2$
32. $y = -x^2 - 8x - 1$ 33. $y = -2x^2 - 8x + 1$ 34. $y = x^2 - 2x + 5$
35. $y = x^2 - 6x + 8$ 36. $y = -x^2 - 2x - 2$ 37. $y = x^2 - x + 2$

38. Find the coordinates of the maximum point of the graph of Equation 1 of Section 36, and thus determine the time for the ball to reach its maximum height and the maximum height it reaches.
39. Show that a rectangle with given perimeter has the largest area when it is a square. *Hint.* Let the perimeter equal $2k$, the length equal x, and the area equal A. What is the width? What is the area, A, in terms of the width and the length?
40. When one cubic foot of coal gas is mixed with x cubic feet of air, the foot-pounds of work done in exploding the mixture is given by the formula $w = 83x - 3.2x^2$. Find the number of cubic feet of air which should be mixed with one cubic foot of coal gas for a maximum amount of work done by exploding the mixture.

37. THE RATE OF CHANGE OF THE QUADRATIC FUNCTION

In Section 31 we saw that the average rate of change of a function is given by the expression

$$\frac{f(x+h) - f(x)}{h} \tag{1}$$

For the quadratic function $f(x) = ax^2 + bx + c$, we have

$$f(x+h) = a(x+h)^2 + b(x+h) + c$$
$$= ax^2 + 2ahx + ah^2 + bx + bh + c$$
$$f(x+h) - f(x) = 2ahx + ah^2 + bh$$
$$\frac{f(x+h) - f(x)}{h} = 2ax + b + ah \tag{2}$$

Equation 2 is, therefore, the expression for the average rate of change of the quadratic function. Unlike the linear function, the average rate of change of the quadratic function is not constant, since it depends on both the value of x and the interval h.

EXAMPLE 1. Find the average rate of change of the function $y = 2x^2 - 3x + 4$ as x changes from $x = 3$ to $x = 4$.

Solution. In Equation 2, $x = 3$ and $h = 1$. Therefore, the average rate of change is $2(2)(3) - 3 + (2)(1) = 11$.

What is a Quadratic Function?

EXAMPLE 2. Find the average rate of change of the function $y = 2x^2 - 8x + 6$ as x changes from $x = 1$ to $x = 1.5$.

Solution: In Equation 2, $x = 1$ and $h = 0.5$. Therefore, the average rate of change is $2(2)(1) - 8 + 2(0.5) = -3$.

A geometric interpretation of the average rate of change will be given later in this section.

The Instantaneous Rate of Change of the Quadratic Function

In Section 31 the instantaneous rate of change of a function was expressed as

$$\operatorname*{Lim}_{h \to 0} \frac{f(x + h) - f(x)}{h} \tag{3}$$

Applying this expression to Equation 2, we have the important result: the instantaneous rate of change of the quadratic function $f(x) = ax^2 + bx + c$ is

$$2ax + b \tag{4}$$

since ah in Equation 2 tends to zero as h tends to zero.

It will be instructive at this point to study the rate of change of a quadratic function by direct evaluation of the function for values of x which are approaching a fixed number. Let us choose the function $y = 2x^2 - 3x + 4$ and evaluate y for the specific values of x in the table.

x	3	3.5	3.4	3.3	3.2	3.1	3.01	3.001
y	13	18	16.92	15.88	14.88	13.92	13.0902	13.009002

To calculate the rate of change for the various intervals we arrange the work in tabular form as follows. (We could, of course, obtain these results directly through the use of Equation 2.)

Interval	Change in x	Change in y	$\dfrac{\text{Change in } y}{\text{Change in } x}$
$x = 3$ to $x = 3.5$	0.5	5	10
$x = 3$ to $x = 3.4$	0.4	3.92	9.8
$x = 3$ to $x = 3.3$	0.3	2.88	9.6
$x = 3$ to $x = 3.2$	0.2	1.88	9.4
$x = 3$ to $x = 3.1$	0.1	0.92	9.2
$x = 3$ to $x = 3.01$	0.01	0.0902	9.02
$x = 3$ to $x = 3.001$	0.001	0.009002	9.002

37. The Rate of Change of the Quadratic Function

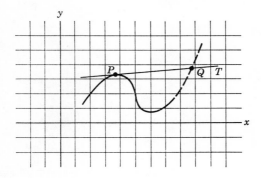

Figure 7.5

Although the computations are laborious, they exhibit convincing evidence that the average rate of change, as tabulated in the last column, approaches a limit and that this limit is the instantaneous rate of change at $x = 3$.

The Tangent to a Curve

The definition of a tangent to a circle, stated in Section 23, does not apply to tangents to other curves. In Figure 7.5, if we consider only the solid part of the curve, tangent PT touches the curve in only one point, P. If, however, we consider the entire curve, tangent PT intersects the curve in another point, Q.

We may remedy this situation by stating that PT is tangent to the curve *at point P* and by redefining a tangent to a curve in the following manner. In Figure 7.6 the line l which intersects the curve in $P_1(x_1, y_1)$ and $P_2(x_2, y_2)$ is called a secant, as in the case of a line through two distinct points of a circle. In Figure 7.7 we maintain point P_1 as a fixed point but allow point P_2 to move

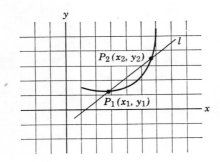

Figure 7.6

224   What is a Quadratic Function?

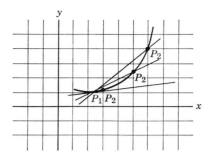

Figure 7.7

down the curve toward point P_1. In general, this secant will approach a limiting position as P_2 tends to coincide with P_1. We call this limiting position of the secant a *tangent* to the curve at P_1 and state the following definition.

Definition 2. The tangent to a curve at a point P_1 on a curve is the limiting position of a secant $P_1 P_2$ as P_2 moves along the curve and tends to coincide with P_1.

A Geometric Interpretation of the Rate of Change of the Quadratic Function

In Figure 7.8 we see that the slope of the secant $P_1 P_2$ is given by the expression for the average rate of change, that is,

$$\text{slope of secant} = \frac{f(x+h) - f(x)}{h} = 2ax + b + ah \tag{5}$$

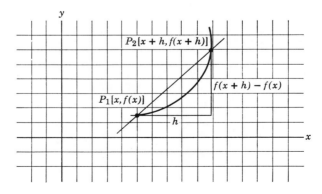

Figure 7.8

37. The Rate of Change of the Quadratic Function

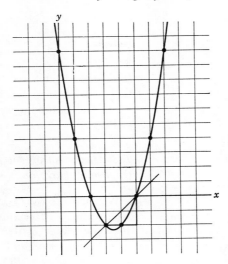

Figure 7.9

EXAMPLE 3. Find the slope of the secant of the parabola $y = x^2 - 7x + 10$ which passes through the points $(3, -2)$ and $(5, 0)$.

Solution: From the given data, $h = 2$. The slope of the secant is then $2(1)(3) - 7 + 1(2) = 1$. We may verify the result from the graph of the parabola, Figure 7.9.

In Figure 7.8 the tangent at P_1 is obtained by moving P_2 along the curve toward P_1, that is, by letting h approach zero. But this is precisely the definition of the instantaneous rate of change, that is,

$$\text{slope of tangent} = \lim_{h \to 0} \frac{f(x + h) - f(x)}{h} = 2ax + b \qquad (6)$$

We then have this very important result:

> The slope of the tangent to the parabola $y = ax^2 + bx + c$, at the point $P(x, y)$, is equal to $2ax + b$.

We are now in a position to find the equation of the tangent to a parabola at a given point.

EXAMPLE 4. Find the equation of the tangent to the parabola $y = 2x^2 - x - 3$ at the point $(2, 3)$.

Solution: By Equation 6 the slope of the tangent is $2(2)(2) - 1 = 7$. Since the equation of any straight line is $y = mx + b$, we may write the

226   What is a Quadratic Function?

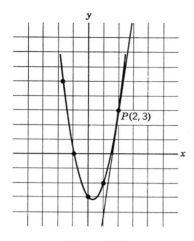

Figure 7.10

equation of the tangent in the form

$$y = 7x + b$$

Since $y = 3$ when $x = 2$, we find that $b = -11$. Therefore, the equation of the tangent is $y = 7x - 11$. Figure 7.10 illustrates the parabola and the tangent line at (2, 3).

The Maximum or Minimum Value of the Quadratic Function

We shall use the results of the preceding discussion to determine the maximum or minimum value of the quadratic function in a manner different from the method used in Section 36. Our method will be intuitive but highly plausible. From Figure 7.11 we infer that at a maximum or minimum of a parabola the tangent line is horizontal, that is, the slope is zero. We may then state that at a maximum or minimum value of the quadratic function,

$$2ax + b = 0$$
$$x = -\frac{b}{2a}$$

as obtained in Section 36.

Physical Interpretation of the Rate of Change of the Quadratic Function

A very meaningful interpretation of the rate of change of the quadratic function is obtained when we study the motion of a projectile shot vertically

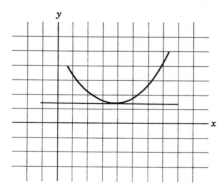 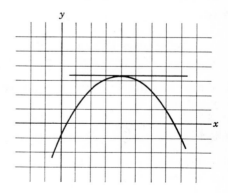

Figure 7.11

upward or downward. The approximate functional representation of this motion is expressed by the quadratic function

$$s = -\tfrac{1}{2}gt^2 + v_0 t + h \tag{7}$$

where
- s = distance, in feet, above the ground after t seconds
- g = acceleration of gravity (approximately -32 feet per second^2)
- t = time in seconds
- v_0 = initial velocity in feet per second (positive if upward, negative if downward)
- h = height, in feet, above the ground at $t = 0$

EXAMPLE 5. A bullet is fired straight upward from a tower 250 feet high with a muzzle velocity of 1600 feet per second. Find:

(a) The function expressing the motion of the bullet.
(b) The velocity at any time t.
(c) The velocity at the end of 10 seconds,
(d) The velocity at the end of 60 seconds; interpret the sign.
(e) The time to reach the maximum height.
(f) The maximum height attained.

Solution:

(a) From Equation 7, $s = -16t^2 + 1600t + 250$.
(b) Since velocity is the rate of change of distance with respect to time, from Equation 4 we have $v = -32t + 1600$.
(c) For $t = 10$, $v = -320 + 1600 = 1280$ feet per second.
(d) For $t = 60$, $v = -1920 + 1600 = -320$ feet per second. The negative sign indicates that the bullet has reached its maximum height and is going down.
(e) Since the maximum height is reached when the velocity is zero, $-32t + 1600 = 0$, or $t = 50$ seconds. (We may, of course, use the formula $t = -(b/2a)$, with the same result.)
(f) For $t = 50$, $s = -16(50)^2 + 1600(50) + 250$, or $s = 40{,}250$ feet.

EXERCISE 37

THE RATE OF CHANGE OF THE QUADRATIC FUNCTION

Find the average rate of change of each of the following functions in the given interval.

1. $f(x) = x^2 - 6x + 3$, $x = 1$ to $x = 2$
2. $f(x) = x^2 - 4x + 1$, $x = 2$ to $x = 3$

3. $f(x) = x^2 - 2x + 2$, $x = 1$ to $x = 3$
4. $f(x) = 2x^2 - 2x + 3$, $x = 3$ to $x = 4$
5. $f(x) = x^2 - 2x - 10$, $x = -1$ to $x = 0$
6. $f(x) = -x^2 - 4x + 4$, $x = 1$ to $x = 1.5$
7. $f(x) = x^2 - 4x - 5$, $x = -2$ to $x = -1$
8. $f(x) = 2x^2 + 11x - 6$, $x = -3$ to $x = -2$

Find the instantaneous rate of change of each of the following functions at the indicated point.

9. $f(x) = x^2 - 2x + 1$, $(1, 0)$
10. $f(x) = x^2 - 4x + 6$, $(2, 2)$
11. $f(x) = 2x^2 - 3x + 2$, $(2, 4)$
12. $f(x) = 3x^2 - 6x + 5$, $(1, 2)$
13. $f(x) = -x^2 + 4x + 4$, $(-2, -8)$
14. $f(x) = x^2 - 4x + 8$, $(2, 4)$
15. $f(x) = x^2 - 6x$, $(0, 0)$
16. $f(x) = x^2 + 9$, $(0, 9)$

Find the equation of the tangent to each of the following parabolas at the given point. Sketch each parabola and the tangent as determined by its equation.

17. $y = x^2 - 4x + 3$, $(1, 0)$
18. $y = x^2 + 4x - 4$, $(2, 8)$
19. $y = x^2 - 6x + 4$, $(1, -1)$
20. $y = -x^2 + 4x + 4$, $(-1, -1)$
21. $y = -x^2 + 4x - 4$, $(1, -1)$
22. $y = x^2 + 6x$, $(-1, -5)$

Find the point on each of the following parabolas where the tangent has the indicated slope.

23. $f(x) = x^2 - 6x + 4$, slope $= 2$
24. $f(x) = x^2 + 6x + 3$, slope $= 1$
25. $f(x) = 2x^2 - 5x + 3$, slope $= 7$
26. $f(x) = x^2 - 4x + 6$, slope $= -2/3$
27. The equation of motion of a projectile shot vertically upward is given by the function $s = -16t^2 + 160t + 40$. Discuss the motion in terms of the information required in Example 5.
28. The equation of motion of a projectile shot vertically upward from the ground is $s = -16t^2 + 160t$. Discuss the motion in terms of the information required in Example 5.

38. QUADRATIC EQUATIONS AND THEIR SOLUTIONS

The zeros of the quadratic function

$$y = ax^2 + bx + c \qquad (1)$$

are determined from the related *quadratic equation*

$$ax^2 + bx + c = 0 \qquad (2)$$

since the zeros of Equation 1 are those values of x for which $y = 0$. Each of the following equations are quadratic equations, since x appears to the second power and no higher.

$$x^2 - 3x + 5 = 0 \tag{3}$$
$$-3x^2 - 2x = 7 \tag{4}$$
$$-2x^2 - 8 = 0 \tag{5}$$
$$-3x^2 = 9x \tag{6}$$
$$6x^2 = 0 \tag{7}$$

When the quadratic equation is written with $a > 0$ and the right-hand member equal to zero, as in Equation 3, it is said to be in *standard form*. Obviously Equations 4, 5, and 6 may be rewritten in standard form by applying the axioms of equality discussed in Section 16. Thus, Equations 4, 5, and 6 become

$$3x^2 + 2x + 7 = 0 \tag{8}$$
$$2x^2 + 8 = 0 \tag{9}$$
$$3x^2 + 9x = 0 \tag{10}$$

Solution of the Quadratic Equation

The quadratic Equation 2 may be solved in a variety of ways. We may draw the graph of the related Function 1 and estimate the zeros; we may reduce the equation to two linear equations by factoring (when convenient), and we may develop a formula which is applicable to any quadratic equation.

Solution by Graph

Consider the quadratic equation

$$4x^2 + 5x - 6 = 0 \tag{11}$$

and its related functions

$$y = 4x^2 + 5x - 6 \tag{12}$$

The graph of Equation 12 is drawn in Figure 7.12. From the graph we estimate the zeros to be $3/4$ and -2, which are, therefore, the roots of Equation 11. Substitution of both $3/4$ and -2 for x in Equation 11 verifies that they are the correct roots, since when

$$x = \tfrac{3}{4}, \ 4(\tfrac{3}{4})^2 + 5(\tfrac{3}{4}) - 6 = 0$$

and when

$$x = -2, \ 4(-2)^2 + 5(-2) - 6 = 0$$

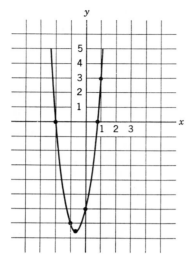

Figure 7.12

While we have correctly estimated the roots in this equation, it should be emphasized that roots obtained graphically are, in general, not exact.

Solution by Factoring

The solution of the quadratic equation by factoring may be made to depend on the solution of two related linear equations. Consider the equation

$$x^2 - x - 42 = 0 \qquad (13)$$

In factored form, Equation 13 becomes.

$$(x - 7)(x + 6) = 0 \qquad (14)$$

We apply a very reasonable assumption.

Axiom 1. The product of two or more factors equals zero if and only if one or more of the factors equal zero.

We may now write the two related linear equations

$$\begin{aligned} x - 7 &= 0 \\ x + 6 &= 0 \end{aligned} \qquad (15)$$

The solutions of Equations 15 are $x = 7$ and $x = -6$. If these values of x are substituted in Equation 13, it will be found that they both satisfy the equation; hence they are the roots of the equation. We give another example of the solution of a quadratic equation by factoring.

EXAMPLE 1. Solve the equation $x^2 = 8x - 15$.
Solution:

$$x^2 - 8x + 15 = 0$$
$$(x - 3)(x - 5) = 0$$
$$x - 3 = 0 \text{ and } x - 5 = 0$$
$$x = 3 \text{ and } x = 5$$

Substituting the roots, $x = 3$ and $x = 5$, in the given equation will verify that $x = 3$ and $x = 5$ are solutions, or roots, of the given equation. It should be emphasized that the equation must be written in standard form before Axiom 1 can be applied.

Solution by Formula

Although the equation
$$x^2 + 2x - 5 = 0 \qquad (16)$$
may be factored as
$$(x + 1 - \sqrt{6})(x + 1 + \sqrt{6}) \qquad (17)$$
this is a rather difficult procedure by the usual methods of factoring. We therefore seek a more general method of solving the quadratic equation which will apply to all quadratic equations.

In Section 36 the quadratic function was written as
$$y = a\left[\left(x + \frac{b}{2a}\right)^2 - \left(\frac{\sqrt{b^2 - 4ac}}{2a}\right)^2\right] \qquad (18)$$

The quantity inside the brackets is the difference of two squares. Therefore, Equation 18 may be written in factored form as
$$y = a\left(x + \frac{b}{2a} + \frac{\sqrt{b^2 - 4ac}}{2a}\right)\left(x + \frac{b}{2a} - \frac{\sqrt{b^2 - 4ac}}{2a}\right) \qquad (19)$$

If we now let $y = 0$, to form the related quadratic equation, and apply Axiom 1, we have the related linear equations
$$\begin{aligned} x + \frac{b}{2a} + \frac{\sqrt{b^2 - 4ac}}{2a} = 0 \\ x + \frac{b}{2a} - \frac{\sqrt{b^2 - 4ac}}{2a} = 0 \end{aligned} \qquad (20)$$

We may combine the solutions of Equations 20 into one equation,
$$x = \frac{-b \pm \sqrt{b^2 - 4ac}}{2a} \qquad (21)$$

Equation 21 is known as the *quadratic formula*.

Using the Quadratic Formula

To solve a quadratic equation by means of the formula, the following procedure is recommended.

1. Write the equation in standard form.
2. Write the specific values of a, b, and c.
3. Substitute these values in the formula.

4. Simplify the result.
5. Find the two roots by using the plus sign and the minus sign in the numerator.
6. If desired, irrational roots may be evaluated to three decimal places by using the Table of Squares and Square Roots in the appendix.
7. Check the roots found in Step 5 or Step 6.

EXAMPLE 2. Solve the equation $2x^2 - 5x + 3 = 0$ by means of the quadratic formula.
Solution: Following the steps suggested above, we have:

(1) $2x^2 - 5x + 3 = 0$
(2) $a = 2, b = -5, c = 3$
(3) $x = \dfrac{5 \pm \sqrt{(-5)^2 - (4)(2)(3)}}{(2)(2)}$
(4) $x = (5 \pm \sqrt{1})/4$
(5) $x = (5 + 1)/4 = 3/2, x = (5 - 1)/4 = 1$
(6) If $x = 3/2$, and $x = 1$, are substituted in the original equation, it will be found that both values satisfy the equation.

It may be noted that this particular equation could have been easily solved by factoring.

EXAMPLE 3. Solve the equation $x^2 = 4x - 1$.
Solution: Following the steps suggested above, we have:

(1) $x^2 - 4x + 1 = 0$
(2) $a = 1, b = -4, c = 1$
(3) $x = \dfrac{4 \pm \sqrt{(-4)^2 - (4)(1)(1)}}{(2)(1)}$
(4) $x = (4 \pm \sqrt{12})/2$[1]
(5) $x = (4 + \sqrt{12})/2, x = (4 - \sqrt{12})/2$
(6) From the table in the Appendix, we find that $\sqrt{12} = 3.464$. The roots are then $(4 + 3.464)/2 = 3.732$, and $(4 - 3.464)/2 = 0.268$.

---

[1] Since $\sqrt{12} = 2\sqrt{3}$, this answer is not in simplest form. However, if the roots are to be expressed in decimal form, the roots as expressed in Step 5 are just as easy to calculate as they are in the simplified form. The student should consult any standard text on algebra for further details on the simplification of radicals.

(7) The equation may be checked with these values of the roots, but the student should bear in mind that since these values are approximations, the check may not be exact.

EXERCISE 38

Quadratic Equations and Their Solutions

Estimate the roots of each of the following quadratic equations from the graph of the related function.

1. $x^2 - 6x + 5 = 0$
2. $x^2 - x - 20 = 0$
3. $x^2 - 7x + 4 = 0$
4. $9x^2 - 30x = -25$
5. $2x^2 - 5x + 3 = 0$
6. $x^2 - x = 3$
7. $x^2 - 3x = 2$
8. $x^2 - 3x - 1 = 0$
9. $x^2 - x + 2 = 0$

Solve each of the following quadratic equations by factoring.

10. $x^2 - 16 = 0$
11. $x^2 - 64 = 0$
12. $x^2 - 6x = 0$
13. $x^2 - 8x = 0$
14. $x^2 - 11x - 42 = 0$
15. $x^2 - x = 90$
16. $x^2 = 5x - 4$
17. $2x^2 - 7x = 15$
18. $2x^2 + x - 15 = 0$
19. $6x^2 = 13x + 5$
20. $8x^2 = -6x + 9$
21. $4x^2 - 12x + 9 = 0$

Solve each of the following quadratic equations by means of the quadratic formula. Evaluate irrational roots to three decimal places, as in Example 3.

22. $2x^2 - 5x + 3 = 0$
23. $3x^2 + 5x - 2 = 0$
24. $2x^2 = 10 - x$
25. $4x^2 - 8x + 3 = 0$
26. $6x^2 - 13x = -6$
27. $x^2 + x - 1 = 0$
28. $2x^2 + 4x - 3 = 0$
29. $3x^2 + 6x + 2 = 0$
30. $3x^2 - 8x + 2 = 0$
31. $2x^2 + 6x + 1 = 0$
32. $4x^2 - 3x - 2 = 0$
33. $4x^2 + 3x - 2 = 0$
34. Express the quadratic equation $ax^2 + bx + c = 0$ in set notation.
35. Express Axiom 1 as a bi-conditional.
36. Express Axiom 1 in terms of necessary and sufficient conditions.

39. EXTENDING THE USE OF THE QUADRATIC FORMULA

The quadratic formula

$$x = \frac{-b \pm \sqrt{b^2 - 4ac}}{2a} \qquad (1)$$

was developed in Section 38 as an aid in the solution of nonfactorable quadratic equations. It applies, of course, to all quadratic equations, and we may regard Equation 1 as a general expression for the roots of any quadratic equation. This generalized solution of the quadratic equation gives us the additional advantage of being able to predict certain characteristics of the roots of the equation and certain properties of the related function. It also furnishes a rather simple check on the accuracy of the solution and enables us to solve a number of problems in an elegant manner, as we shall see later in this section.

The Sum and Product of the Roots

If the quadratic formula is written in the form

$$x_1 = \frac{-b + \sqrt{b^2 - 4ac}}{2a}$$

$$x_2 = \frac{-b - \sqrt{b^2 - 4ac}}{2a}$$

where x_1 and x_2 are the roots of the equation, we may readily derive two important relations between the roots and coefficients of the equation. These relations are

$$x_1 + x_2 = -\frac{b}{a}$$

$$x_1 \cdot x_2 = \frac{c}{a} \tag{2}$$

These relations may be more easily recalled if we write the equation in the form

$$x^2 + \frac{b}{a}x + \frac{c}{a} = 0 \tag{3}$$

Thus, when the coefficient of x^2 in the quadratic equation equals 1, *the sum of the roots equals the negative of the coefficient of x, and the product of the roots equals the constant term.*

EXAMPLE 1. Find the sum and product of the roots of the equation $3x^2 - 2x + 1 = 0$.

Solution: We first write the equation in the form

$$x^2 - \tfrac{2}{3}x + \tfrac{1}{3} = 0$$

We have at once

$$x_1 + x_2 = -(-\tfrac{2}{3}) = \tfrac{2}{3}$$

$$x_1 \cdot x_2 = \tfrac{1}{3}$$

Checking the Roots of a Quadratic Equation

While the relations in Equations 2 do not lead to solution of the quadratic equation, as the student will be asked to show in Problem 50 of this section, they do provide a simple method of checking the roots of the equation.

EXAMPLE 2. Solve and check the roots of the equation $2x^2 - 3x + 1 = 0$.
Solution: $(2x - 1)(x - 1) = 0$

$$x = \tfrac{1}{2}$$
$$x = 1$$

Check. We write the equation as

$$x^2 - \tfrac{3}{2}x + \tfrac{1}{2} = 0$$

$\tfrac{1}{2} + 1 = \tfrac{3}{2}$, the negative of the coefficient of x
$\tfrac{1}{2}(1) = \tfrac{1}{2}$, the constant term

Forming a Quadratic Equation with Given Roots

While a quadratic equation may be determined from its given roots by reversing the process of the solution by factoring, the relations in Equations 2 are very convenient for this purpose. We shall show examples of each method.

EXAMPLE 3. Determine the quadratic equation whose roots are $1/2$ and $-(2/3)$.
Solution: Method 1. $x = \tfrac{1}{2}$ $x = -\tfrac{2}{3}$

$$x - \tfrac{1}{2} = 0 \qquad x + \tfrac{2}{3} = 0$$
$$(x - \tfrac{1}{2})(x + \tfrac{2}{3}) = 0$$
$$(2x - 1)(3x + 2) = 0$$
$$6x^2 + x - 2 = 0$$

Method 2. From the sum and product relations, we have

$$\frac{b}{a} = -(\tfrac{1}{2} - \tfrac{2}{3}) = \tfrac{1}{6}$$

$$\frac{c}{a} = (\tfrac{1}{2})(-\tfrac{2}{3}) = -\tfrac{1}{3}$$

Then
$$x^2 + \tfrac{1}{6}x - \tfrac{1}{3} = 0$$

$6x^2 + x - 2 = 0$, which is the required equation

EXAMPLE 4. Determine the quadratic equation whose roots are $x = 2 + \sqrt{3}$ and $x = 2 - \sqrt{3}$.

Solution: The second method is more convenient in this example:

$$\frac{b}{a} = -(2 + \sqrt{3} + 2 - \sqrt{3}) = -4$$

$$\frac{c}{a} = (2 + \sqrt{3})(2 - \sqrt{3}) = 1$$

Then $\qquad x^2 - 4x + 1 = 0,$ which is the required equation

The Discriminant $b^2 - 4ac$

The quantity $b^2 - 4ac$ in the quadratic formula is called the discriminant of the equation, since it enables us to determine the nature of the roots of the equation without actually solving it. With it we can also determine certain properties of the graph without actually constructing the graph. By considering the role of the quantity $\sqrt{b^2 - 4ac}$ in the formula, we make the following observations.

In the quadratic equation $ax^2 + bx + c = 0$:

1. If $b^2 - 4ac > 0$, the roots are real and unequal. If $b^2 - 4ac$ is a perfect square, the roots are rational and unequal.
2. If $b^2 - 4ac = 0$, the roots are rational and equal.
3. If $b^2 - 4ac < 0$, the roots fall outside the set of real numbers. In this case, $\sqrt{b^2 - 4ac}$ is called an *imaginary number*, and an expression such as $2 + \sqrt{-3}$ is called a complex number. Therefore, if $b^2 - 4ac < 0$, the roots are complex numbers.

In the quadratic function $y = ax^2 + bx + c$:

1. If $b^2 - 4ac > 0$, the graph intersects the x-axis in two real points.
2. If $b^2 - 4ac = 0$, the graph touches the x-axis in one real point. The zeros of the function are therefore coincident points, and the parabola is said to be tangent to the x-axis.
3. If $b^2 - 4ac < 0$, the graph does not touch nor cross the x-axis.

EXAMPLE 1. Determine the character of the roots of the equation $x^2 - 3x + 2 = 0$, and draw the graph of the related function.

Solution: Since $b^2 - 4ac = 9 - 8 = 1$, the roots are real, unequal, and rational. The graph is shown in Figure 7.13.

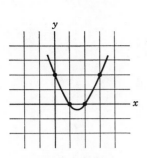
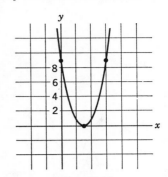

Figure 7.13 **Figure 7.14**

EXAMPLE 2. Determine the character of the roots of the equation $4x^2 - 12x + 9 = 0$, and draw the graph of the related function.

Solution: Since $b^2 - 4ac = 144 - 144 = 0$, the roots are real, equal, and rational. The graph is shown in Figure 7.14.

EXAMPLE 3. Determine the character of the roots of the equation $2x^2 - x + 5 = 0$.

Solution: Since $b^2 - 4ac = 1 - 40 = -39$, the roots are complex numbers. The graph is shown in Figure 7.15.

Application of the Discriminant to a Tangent Problem

Suppose that we wish to determine the equation of a tangent to the parabola $y = 3x^2 - 2x + 5$ with a slope equal to 4. Although we could adapt the method of Section 37 to the solution of this problem, we will demonstrate an alternative method based on properties of the discriminant. Let us consider the pair of equations

$$y = 3x^2 - 2x + 5 \qquad (4)$$
$$y = 4x + k \qquad (5)$$

Equation 4 is the equation of the given parabola, and Equation 5 is the equation of any straight line with slope equal to 4. As in the case of a pair of linear equations, the substitution of Equation 5 in Equation 4 results in an equation with a single variable, x. The solution of this equation gives us the points of intersection of the parabola and the straight line. If we substitute Equation 5 in Equation 4, we have

$$4x + k = 3x^2 - 2x + 5$$

or
$$3x^2 - 6x + 5 - k = 0 \qquad (6)$$

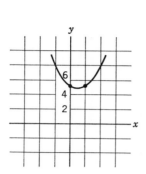

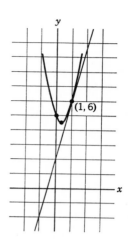

Figure 7.15 **Figure 7.16**

Now, in general, Equation 6 will have two distinct solutions, which represent the two points of intersection of the straight line and parabola. But if we set $b^2 - 4ac$ equal to zero, the two points of intersection will coincide, since the two roots of Equation 6 will be equal. This, of course, is precisely the condition that the straight line (5) be a tangent to the parabola. Therefore, setting $b^2 - 4ac$ in Equation 6 equal to zero, we have

$$36 - 12(5 - k) = 0$$
$$k = 2 \tag{7}$$

If we substitute $k = 2$ in Equation 5 we have

$$y = 4x + 2 \tag{8}$$

which is the equation of the tangent. Furthermore, we can now find the point of tangency. For $k = 2$, Equation 6 becomes

$$3x^2 - 6x + 3 = 0$$

or
$$x^2 - 2x + 1 = 0 \tag{9}$$

Equation 9 has two equal roots, each equal to 1. For $x = 1$, we have from Equation 8, $y = 6$. Therefore, the point of tangency is (1, 6). Figure 7.16 shows the graph of the parabola and the tangent and confirms the correctness of our solution.

39. Extending the Use of the Quadratic Formula

EXERCISE 39
Extending the Use of the Quadratic Formula

Find the sum and product of the roots of each of the following equations.

1. $x^2 - 3x + 2 = 0$
2. $2x^2 - 3x - 2 = 0$
3. $3x^2 - 7x + 7 = 0$
4. $2x^2 + 3x = 1$
5. $x^2 - 7x = 5$
6. $6x^2 - 7x + 2 = 0$
7. $3x^2 - x = 0$
8. $2x^2 - 17 = 0$
9. $3x^2 + 2x - k = 0$

Solve each of the following equations, and check the roots by means of the sum and product relations. (Use decimal approximations of irrational roots.)

10. $x^2 - 4x - 21 = 0$
11. $6x^2 - 7x + 2 = 0$
12. $12x^2 - 71x - 6 = 0$
13. $3x^2 - 8x - 7 = 0$
14. $x^2 - 7x + 4 = 0$
15. $4x^2 + 4x - 1 = 0$
16. $x^2 + x - 3 = 0$

Form the equations that have the given roots.

17. $2, -3$
18. $1/2, 2/3$
19. $1/2, -2/3$
20. $3/5, 4/5$
21. $2/3, -1/3$
22. $2/3, 3/5$
23. $3 + \sqrt{5}, 3 - \sqrt{5}$
24. $-2 + \sqrt{6}, -2 - \sqrt{6}$
25. $5 + \sqrt{2}, 5 - \sqrt{2}$
26. $1.7, -0.7$
27. $2.3, -0.3$

Determine the character of the roots of each of the following equations.

28. $2x^2 - 7x + 3 = 0$
29. $2x^2 - 4x + 3 = 0$
30. $x^2 + 6x - 8 = 0$
31. $3x^2 + 15x - 19 = 0$
32. $6x^2 + x + 1 = 0$
33. $x^2 - 4x + 4 = 0$
34. $x^2 - 2x + 1 = 0$
35. $2x^2 - 3x + 5 = 0$
36. $4x^2 - 20x + 25 = 0$
37. $x^2 + x + 2 = 0$
38. $9x^2 - 12x + 4 = 0$
39. $x^2 - x = 0$

40. Determine the value of k so that the equation $4x^2 + 20x + k$ has equal roots.
41. Determine the value of k so that the equation $x^2 + kx + 4$ has equal roots.
42. Determine the value of k so that the equation $x^2 + 3kx + k + 7$ has equal roots.

Find the equation of the tangent, with given slope, to each of the following parabolas. Draw the graph, indicating the point of tangency.

43. $y = x^2 - 5x + 6$, slope of tangent equals 3
44. $y = 2x^2 - 3x + 5$, slope of tangent equals 5
45. $y = x^2 - 3x + 4$, slope of tangent equals 1
46. $y = x^2 - 4x + 4$, slope of tangent equals 2
47. $y = x^2 - 2x + 3$, slope of tangent equals -1
48. $y = x^2 - 3x + 4$, slope of tangent equals $1/2$

49. $y = 2x^2 - 4x + 1$, slope of tangent equals -2

50. Show that the sum and product relations, Equations 2, do not lead to an easier solution of the quadratic equation. *Hint.* Write Equations 2 in the form $x_1 = -x^2 - (b/a)$, $x_1 = c/ax_2$, then equate values of x_1 and simplify the result.

40. APPLICATIONS OF THE QUADRATIC FUNCTION

In Section 37 we studied the application of the quadratic function to the motion of a projectile. There are many other such applications; we shall discuss two of these and others will be given as problems at the end of this section.

EXAMPLE 1. The velocity of sound in sea water is given by the formula $v = 4756 + 13.8t - 0.12t^2$, where t is the temperature in degrees centigrade, and v is the velocity in feet per second. Find (a) the temperature for the greatest velocity and (b) the temperature when the velocity is 4800 feet per second.

Solution: (a) For maximum velocity (see Equation 6, Section 37)

$$t = -\frac{13.8}{-0.24} = 57.5° \text{ centigrade}$$

Since 57.5° centigrade is equivalent to 135.5° Fahrenheit, it is unlikely that the maximum velocity is ever attained, except under laboratory conditions.

(b) When the velocity is 4800 feet per second,

$$4800 = 4756 + 13.8t - 0.12t^2$$
$$0.12t^2 - 13.8t + 44 = 0$$

Using the quadratic formula, we find the solution to be

$$t = 112.1° \text{ centigrade}$$
$$t = 3.3° \text{ centigrade}$$

Since 112.1° centigrade is above the boiling point, we must discard this solution.

Our second example describes a hypothetical situation which affords an excellent illustration of the analysis of a problem to determine the maximum or minimum value of a function.

EXAMPLE 2. Two straight railroad tracks cross at right angles to each other at a point O. A train starts at a station A on the first track, 10 miles from O,

40. Applications of the Quadratic Function

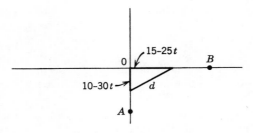

Figure 7.17

traveling toward O at a rate of 30 miles per hour. At the same time a second train starts from a point B on the second track, 15 miles from O, traveling toward O at a rate of 25 miles per hour. When will the distance between the two trains be a minimum?

Solution: From the formula

$$\text{distance} = \text{rate} \times \text{time}$$

at the end of t hours the first train travels $30t$ miles, and the second train travels $25t$ miles. Then the distance d between the two trains is the hypotenuse of the right triangle in Figure 7.17, whose sides are $15 - 25t$ and $10 - 30t$. From the Pythagorean theorem, we then have

$$d^2 = (10 - 30t)^2 + (15 - 25t)^2$$
$$d^2 = 1525t^2 - 1350t + 325$$

The minimum value of d^2 occurs when $t = 1350/3050$, or $27/61$. (When d^2 is a minimum, so is d.) This is approximately 27 minutes after the trains started.

EXERCISE 40

APPLICATIONS OF THE QUADRATIC FUNCTION

1. The greatest distance, d miles, an airplane pilot can see when at a height h miles above the ground is approximately given by the formula $d^2 = h^2 + 800h$. Find the height to which the plane must rise in order for the pilot to see a distance of 10 miles.

2. An approximate formula for the relation between the distance, d feet, for which a car traveling at a rate of r miles per hour can be stopped is $d = 0.045r^2 + 1.1r$. Find the maximum speed a car can travel in order for the driver to avoid hitting an object 40 feet ahead.

3. When the length l and the width w of a rectangle are related by the equation $l^2 - lw - w^2 = 0$, the proportions are considered to be most pleasing to the eye. Find the width of a table top of these proportions if the length is 108 inches.

4. The side s of a regular decagon inscribed in a circle of radius r is given by the formula $s^2 + rs - r^2 = 0$. Find the radius of a circle required for an inscribed decagon with a side equal to 8 inches.

5. When water flows from a reservoir through a pipe 4 inches in diameter and 10 feet in length, the velocity of the water in feet per second is given by the formula $4v^2 + 5v - 2 = 480h$, where h is the height of the reservoir in feet. Find the velocity of the water when the height is 15 feet.

6. A length of tin 20 inches wide is formed into an open gutter with a rectangular cross section. Find the dimensions for greatest carrying capacity.

7. A manufacturer of flashlights finds that if he sells x items his profit is given by the formula $P = 150x - 0.01x^2 - 20{,}000$. How many items should he produce and sell for maximum profit?

8. A camera manufacturer estimates that if he charges $100 for his best model he can sell 200 cameras per month, but he can sell 50 more per month for each $10 reduction in price. What should he charge for the greatest monthly income? Does the answer necessarily indicate the price for greatest profit?

9. The power, P, delivered by a storage battery is given by the formula $P = EI - I^2R$, where E is the fixed electromotive force in volts, R is the fixed internal resistance in ohms, and I is the current in amperes. For what current will the output be a maximum?

10. The total surface area S of a right circular cylinder of radius r and height h is given by the formula $S = 2\pi rh + 2\pi r^2$. Find the radius of a cylinder with a total surface of 192π cubic inches and a height of 10 inches.

11. The strength S of a rectangular beam cut from a log of diameter a is a maximum when the width w is such that $S = a^2w - w^3$. Choosing the diameter a to equal $\sqrt{3}$, carefully graph the function from $w = 0$ to $w = 2$ and estimate the maximum value of S from the graph.

12. An approximate formula for the number of cars, n, passing a given point on a highway during a particular hour is $n = 3(42s - s^2)$, where s is the speed in miles per hour. Find the speed for which the number of cars will be a maximum.

How are Statistics Used to Present and Measure Data?

41. USES OF STATISTICS

If one wishes to find the time for one oscillation of a simple pendulum of given length, he has merely to use a formula of the form

$$T = 0.32\sqrt{L} \tag{1}$$

where T is the time in seconds and L is the length of the pendulum in inches. Formula 1 may be deduced without experimentation on the basis of certain principles of physics, and it applies to all simple pendulums. The important point we wish to emphasize is that while observation might have suggested the formula, it may be established by purely deductive methods.[1] On the other hand, if one wishes to predict the number of entering freshmen in a certain college who will wear glasses, an approximate formula could be determined only by experimentation. What kind of experimentation do we refer to? Over a period of years records would be kept showing the total freshman enrollments and the number of these freshmen wearing glasses. If these data, or *statistics*, showed that, during the period of observation,

---

[1] The constant, 0.32, is determined experimentally.

about two out of every seven freshmen wore glasses, one could reasonably predict that in an incoming class of 1400 freshmen, about 400 would wear glasses. Whether or not the formula suggested by the statistical evidence

$$G = \tfrac{2}{7}F \qquad (2)$$

where G is the number of freshmen wearing glasses and F is the total number of freshmen, would apply to other colleges, would be questionable. In fact it might not apply to the given college in subsequent years; being inductively arrived at, it has the shortcomings discussed in Section 1.

In spite of such limitations, the conclusions based on statistical evidence may carry considerable weight, as in the case of the contention that cigarette smoking is a major cause of lung cancer or that certain "indicators" portend general changes in the stock market. Even such mundane matters as the description of a winter as a "cold winter" or a "mild winter" can be given only on the basis of data on the daily winter temperatures recorded during the season, in much the same way that the semester performance of a student is summarized by a single grade in the course, based on all of his work during the semester.

In this chapter we will consider the use of statistics as a basis for drawing conclusions. Such study generally involves the collection, organization, analysis, and interpretation of data.

Collection of Data: Sampling

Our study of statistics will be concerned primarily with the last three items, the organization, analysis, and interpretation of data. This is not to minimize the importance of proper methods of collection, but we shall find it convenient to assume that the data has been collected from various sources for us to "work on." At this time, however, we shall discuss one important aspect of the problem of data collection, that is, the problem of sampling.

Suppose that we wish to predict the number of incoming freshmen in all colleges throughout the country who will wear glasses. Since, as we have suggested, we might reasonably hesitate to apply information about a particular college to all colleges, the data would best be obtained from all colleges. With a total college freshmen enrollment of 1,565,564[2] students, it would be impractical, if not impossible, to get data on every college freshman, that is, to conduct a *census*. We would, therefore, select a part, or *sample*, of the total *population* (that is, all college freshmen) as representative of the entire group, or population. Just how representative this sample would be

---

[2] From U.S. Department of Health, Education, and Welfare publication OE 54003-66, Opening Fall Enrollment in Higher Education.

is a major concern of statistics. The usual "poll" before a national election relies on the choice of a representative sample for its accuracy of prediction. A well-known example of a failure to collect a representative sample in a nation-wide poll is the poll conducted by the *Literary Digest* in 1936. On the basis of over 2 million questionnaires, a Republican victory was predicted. In the election which followed soon after the results were predicted, President Roosevelt, the Democratic candidate, won by a great majority. One of the reasons for the failure of the poll was that the method of sampling excluded certain groups of people, who tended to vote Democratic, from participating. An almost immediate result of the debacle was the demise of the *Literary Digest*; it could not survive the effects of the mistake it had made.

While, theoretically, errors such as this might be avoided by taking data from the entire population (assuming that everyone questioned would respond, which is far from the case), in most cases this is not only impractical but impossible. A manufacturer who wished to get data, perhaps for advertising purposes, on the life of a flashlight battery, might conceivably test every battery produced. The data would be quite accurate, but the manufacturer would be left without any batteries to sell! Sampling theory would be applied in order to determine the sample needed to pass judgment on all batteries. A relatively large sample is not, of itself, an assurance of an adequate or representative sample; clearly the omission or inclusion of one defective machine or one inefficient employee engaged in making the batteries, would also seriously affect the accuracy of the sample. The failure of the *Literary Digest* was not a result of the size of the sample, which was deemed quite adequate, but rather of the method of selecting the individuals composing the sample. If the batteries made by one inefficient worker were not included in the sample, the tests would indicate a lesser percent of poor batteries than were produced by the factory as a whole.

EXERCISE 41

The Uses of Statistics

In Problems 1–8, state whether a census or a sample would be best used in collecting data on the information sought.

1. A manufacturer of appliances seeks to determine the color preference of purchasers of electric toasters.
2. A Junior College wants to know the four-year colleges to which its students transfer.
3. The Association of American Colleges wants information concerning the ages of students in American Colleges.

4. The makers of fluorescent lighting equipment want data on the life of 15-watt fluorescent lamps.
5. A small elementary school wants to know the number of immunization shots its first-grade pupils have had before entering school.
6. General Motors Corporation wants to know the style preference of car owners.
7. A large ice cream company wants to know the flavor preference of children.
8. A kindergarten teacher wants to know the kinds of ice cream the children want for their Christmas party.

Explain why samples chosen as described in Problems 9–14 might not be reliable.

9. To get data concerning family size, an interviewer selects an area with a large number of high-rise apartments in order to get his information as quickly as he can.
10. To predict the winner of the 1969 baseball pennant, a newspaper conducts a poll in its own city.
11. To estimate the income of the 1967 graduates of a certain college, a survey bases its conclusions on replies received in questionnaires sent to the graduates.
12. A large daily newspaper polls only its subscribers in a presidential poll.
13. To determine the proportion of imperfect screws produced by a factory, a sample of 10,000 screws is taken from one of ten machines.
14. To get data on unemployment in a large city, an interviewer includes people emerging from a large movie theater after an afternoon showing.

42. ORGANIZATION OF DATA

When the data concerning a particular investigation are not extensive and are not to be subjected to mathematical analysis, organization and presentation are a simple matter. Table 8.1 shows the enrollment of an urban junior college over a period of five years. This table clearly shows the enrollment trend for the given period and may form the basis for predicting the enrollment for the next two or three years.

Table 8.1 Fall Enrollment of an Urban Junior College for the Period 1964–1968

Year	Enrollment
1964	33,291
1965	36,478
1966	34,505
1967	36,226
1968	37,224

The Bar Graph

There are several methods of exhibiting the data of Table 8.1 graphically, but we shall discuss only one of them, the *bar graph*. The bars may be drawn either vertically or horizontally, but the comparison of the yearly enrollments is best emphasized by the vertical arrangement. The use of vertical bars is recommended when time is one of the variables. As in all graphs, the title should convey the subject of the graph, and the scale should be clearly indicated. Figure 8.1 is the bar graph for Table 8.1.

Frequency Distributions

The data for most statistical investigations are not as simple as that of Table 8.1, and organization becomes imperative. As an example, consider the following data on the scores of 100 students on a mathematics test. We may assume that the scores have been copied directly from the test papers, arranged in alphabetical order.

31, 42, 48, 35, 45, 49, 33, 44, 47, 58, 59, 56, 21, 35, 38, 55, 46, 37, 36, 32, 30, 24, 60, 52, 48, 47, 39, 26, 36, 44, 47, 54, 58, 51, 46, 41, 39, 21, 27, 42, 43, 42, 55, 28, 38, 40, 45, 49, 46, 51, 56, 33, 30, 42, 29, 34, 37, 43, 49, 45, 55, 58, 52, 33, 36, 41, 58, 52, 49, 55, 38, 40, 37, 42, 49, 43, 41, 41, 44, 38, 44, 49, 44, 42, 41, 45, 44, 32, 45, 32, 37, 45, 58, 30, 30, 45, 51, 39, 44, 49,

The unorganized data convey very little information to the instructor, particularly if he wishes to determine the grade corresponding to a particular score. We could improve the situation slightly by ordering the scores from highest to lowest, or vice versa, but there would still be about 40 different scores to list. It serves our purpose much better to organize the data into *classes* (that is, subsets of the entire set of scores) in a *frequency table*, or

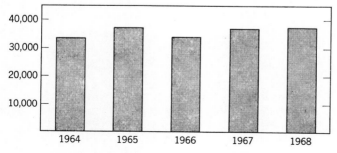

Figure 8.1 Fall enrollment of an urban junior college.

Table 8.2 Scores Made by Students in a Mathematics Test

Class Interval h	Mid-Value x	Tallies	Frequency f
55.5–60.5	58	ⅢⅡ ⅠⅠⅠⅠ	9
50.5–55.5	53	ⅢⅡ ⅢⅡ Ⅰ	11
45.5–50.5	48	ⅢⅡ ⅢⅡ ⅢⅡ	15
40.5–45.5	43	ⅢⅡ ⅢⅡ ⅢⅡ ⅢⅡ ⅢⅡ ⅠⅠⅠ	28
35.5–40.5	38	ⅢⅡ ⅢⅡ ⅢⅡ Ⅰ	16
30.5–35.5	33	ⅢⅡ ⅢⅡ	10
25.5–30.5	28	ⅢⅡ ⅠⅠⅠ	8
20.5–25.5	23	ⅠⅠⅠ	3
Total			100

frequency distribution, as shown in Table 8.2. The measurements, which may be scores on a test, heights of men, inches of rainfall, etc., all called *variates*, and the *frequency* refers to the number of times each variate or class of variates occurs. The frequency distribution not only has the advantage of condensing the data into eight classes but simplifies the assignment of grades. Thus, the table might suggest a score of 55 to 60 for an A, a score of 46 to 54 for a B, and so forth.

The Elements of a Frequency Table

Definition 1. The *class interval*, sometimes called the *width* of the class, is the difference between the boundaries of each class.

In Table 8.2 the class interval is 5. Class intervals are usually of equal width.

Definition 2. The *class limits*, are the boundaries, or endpoints, of each class.

In Table 8.2 the class limits of the first class are 55.5 and 60.5. In general, the class limits are given to one more decimal place than the data, so that no variate will fall into more than one class interval. If the intervals were 55–60, 50–55, and so on, a score of 55 could be put in either the first or the second class.

Definition 3. The *class mark* is the midpoint of each class.

In Table 8.2 the first class mark is 58, the second 53, and so on.

There is no fixed rule for determining the number of classes; it is common practice to have no less than six classes and no more than 20. When the number of classes is small, too many scores may be put into the same class, with a resulting loss of accuracy. This is because the class mark, which is taken as representative of the class, will represent too wide a spread of variates. On the other hand, when too many classes are included, the benefits of condensing the data are lost. We shall see in the next section the effect of varying the number and therefore, the width of the class intervals.

Graphing a Frequency Distribution

As in the case of most relations, the graph may be used to advantage in connection with a frequency distribution. The most commonly used graph is the *histogram*, shown in Figure 8.2. The frequencies are indicated on the vertical scale, and the class limits are indicated on the horizontal scale. Unlike the bar graph, there is no space between the bars.

A *frequency polygon* is a broken-line graph, with the class marks on the horizontal scale. Figure 8.3 is a frequency polygon for the data of Figure 8.2. Note that the frequency polygon may be constructed from a histogram by connecting the midpoints of the top of each bar by straight lines.

The *cumulative frequency polygon*, or *ogive*, as it is commonly called, is a broken-line graph showing the number of frequencies above or below the boundary of each class. Table 8.3 is adapted from Table 8.1 to show "more than" and "less than" cumulative frequencies, and Figure 8.4 shows the ogives for both frequencies. One of the advantages of the cumulative

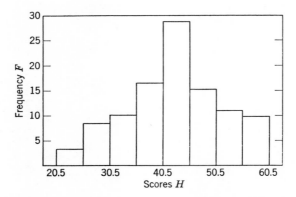

Figure 8.2 Scores made by students in a mathematics test.

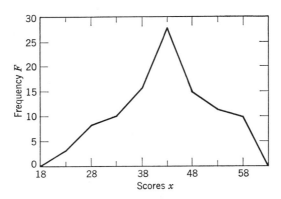

Figure 8.3 Scores made by students in a mathematics test.

Table 8.3 Scores Made by Students in a Mathematics Test

Class Interval h	Mid-Value x	Frequency f	Cumulative Frequency	
			More than	Less than
55.5–60.5	58	9	9	100
50.5–55.5	53	11	20	91
45.5–50.5	48	15	35	80
40.5–45.5	43	28	63	65
35.5–40.5	38	16	79	37
30.5–35.5	33	10	89	21
25.5–30.5	28	8	97	11
20.5–25.5	23	3	100	3

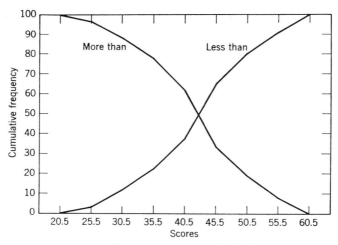

Figure 8.4 Ogive for Table 8.3.

frequency column is that it provides information concerning individual scores in the distribution. A score of 50, for example, may be meaningless as a measure of ability unless one notes that only 20 of the 100 students did better.

It should be noted in Table 8.3 that, in any interval, the "more than" refers to more than the lower limit of the interval and the "less than" refers to less than the upper limit of the interval. Thus, for the interval 40.5–45.5, there are 63 scores more than 40.5, and there are 65 scores less than 45.5.

EXERCISE 42

ORGANIZATION OF DATA

1. Draw a vertical bar graph for the following data.

   Areas of Continents in Square Miles

Continent	Area (to nearest 100,000 square miles)
Africa	11,600,000
Asia	17,000,000
Australia	3,000,000
Europe	3,800,000
North America	9,400,000
South America	6,900,000

2. The following data represent the grades of 51 students on a Mathematics test:

   42, 26, 34, 35, 58, 49, 53, 43, 43, 40, 21, 45, 39, 37, 44, 48, 24, 32, 55, 60, 50, 41, 43, 36, 45, 44, 53, 50, 27, 38, 40, 42, 41, 59, 54, 31, 30, 29, 36, 49, 56, 53, 46, 49, 38, 40, 44, 32, 41, 35, 39.

   (a) Arrange the given data in the form of a frequency distribution. Use the intervals 20.5–25.5, 25.5–30.5, ... 55.5–60.5. Use the form of Table 8.2, Section 41.
   (b) Draw a histogram.
   (c) Draw a frequency polygon.

3. The following data represent the grades of 100 seventh-grade pupils on a spelling test.

100	98	98	55	95	97	96	100	97	95	97	72	97	69	99	99
97	74	99	98	100	99	98	100	87	71	77	93	80	93	98	88
78	76	99	90	60	90	73	78	89	83	100	90	80	99	78	90
98	87	66	80	89	77	70	53	60	94	87	75	94	60	86	91
99	80	78	86	98	89	99	73	83	87	89	98	79	100	97	60
93	89	69	93	90	78	77	89	100	85	78	91	78	46	49	61
72	85	87	89												

(a) Arrange the given data in the form of a frequency distribution. Use the intervals 40.5–45.5, 45.5–50.5, ..., 95.5–100.5. Use the form of Table 8.2, Section 41.
(b) Draw a histogram.
(c) Draw a frequency polygon.
(d) Using the form of Table 8.3, calculate a "more than" and a "less than" cumulative frequency polygon.

4. Repeat Problem 3 using 8 class intervals instead of 12. *Suggestion.* Use 45.5 for the lower-end value, and a class interval of 7.
5. The following data represent scores on an English placement test for college freshmen.

36, 27, 21, 31, 37, 10, 8, 21, 23, 35, 17, 18, 30, 24, 24, 7, 21, 25, 20, 24, 26, 29, 31, 18, 30, 24, 25, 20, 22, 30, 19, 10, 28, 15, 43, 32, 26, 18, 17, 24, 22, 28, 17, 19, 16, 27, 21, 36, 14, 22, 17, 19, 29, 14, 44, 11, 23, 16, 34, 30, 20, 46, 28, 40, 46, 20, 27, 21, 15, 17, 24, 17, 42, 40, 28, 8, 26, 17, 37, 29, 15, 34, 16, 39, 40, 40, 21, 44, 23, 10, 13, 25, 21, 33, 31, 21, 6, 11, 17, 37, 24, 23, 31, 37, 8, 44, 27, 32, 28, 14, 40, 32, 33, 29, 4, 17, 12, 7, 31, 34, 32, 38, 41, 21, 34,

(a) Determine the total number of scores.
(b) Determine the range, that is, the lowest and the highest scores.
(c) Construct a frequency table using 11 class intervals. Check that the total frequency agrees with the frequency found in part *a*.
(d) Draw a frequency polygon.

43. MEASURES OF CENTRAL TENDENCY

In Section 42 we saw that an extensive collection of data could be summarized in a frequency distribution to convey information which would be difficult to obtain otherwise. While Table 8.2, as we have seen, could be useful in assigning grades corresponding to certain groups of marks and in comparing the scores of the students with each other, we need more information if we desire to compare the scores of the 100 students with the scores of other similar groups taking the same test. For this purpose it is desirable to have a single measure characteristic of the group as a whole, just as a single course grade at the end of the semester characterizes the performance of a student for the entire course. We shall discuss three such measures, called measures of central tendency: the *arithmetic mean* or *average*, the *median*, and the *mode*. We shall first introduce certain mathematical symbols useful in defining and calculating these measures.

Some Mathematical Notations

Many of the formulas of statistics may be written in more concise form by means of certain symbols. In particular, the symbol Σ, a capital *sigma* in the

Greek alphabet, is extensively used. The meaning of Σ is defined by the equation

$$\sum_{i=1}^{n} x_i = x_1 + x_2 + \cdots x_n \qquad (3)$$

where, as in Section 27, x_i is any particular value of x, and n is the number of terms. The symbol $\sum_{i=1}^{n} x_i$ is read, "The summation of x_i from $i = 1$ to $i = n$." Thus

$$\sum_{i=1}^{5} x_i = x_1 + x_2 + x_3 + x_4 + x_5$$

A few extensions of this notation will be stated here.

$$\sum_{i=1}^{n} (x_i + y_i - z_i) = \sum_{i=1}^{n} x_i + \sum_{i=1}^{n} y_i - \sum_{i=1}^{n} z_i \qquad (4)$$

$$\sum_{i=1}^{n} kx_i = k \sum_{i=1}^{n} x_i \quad (k \text{ a constant}) \qquad (5)$$

$$\sum_{i=1}^{n} k = nk \quad (k \text{ a constant}) \qquad (6)$$

$$\sum_{i=1}^{n} f_i x_i = f_1 x_1 + f_2 x_2 + \cdots + f_n x_n \qquad (7)$$

EXAMPLE 1.
$$\sum_{i=1}^{4} 5x_i = 5 \sum_{i=1}^{n} x_i = 5(x_1 + x_2 + x_3 + x_4)$$

EXAMPLE 2.
$$\sum_{i=1}^{6} 5 = 6(5) = 30$$

EXAMPLE 3.
$$\sum_{i=1}^{8} (x_i - 5) = \sum_{i=1}^{8} x_i - \sum_{i=1}^{8} 5 = \sum_{i=1}^{n} x_i - 40$$

One more symbol, $|x|$, read "the absolute value of x," will be useful in a later section. The absolute value of a number indicates its size, or magnitude, without regard to its sign. Thus

$$|4| = 4, \qquad |-4| = 4, \qquad |0| = 0$$

A formal definition of absolute value is given by the following relations:

$$|x| = \begin{cases} x & \text{if } x \geq 0 \\ -x & \text{if } x < 0 \end{cases}$$

The Arithmetic Mean of Ungrouped Data

For ungrouped data, the arithmetic mean is defined as the sum of the variates divided by the number of variates. Using the symbol $\bar{x}$ (read "x-bar") for the arithmetic mean, and the symbols $x_1, x_2, \ldots, x_n$ for the n variates, we have:

Definition 4.

$$\bar{x} = \frac{x_1 + x_2 + \cdots + x_n}{n} \tag{8}$$

or, in the sigma notation

$$\bar{x} = \frac{\sum_{i=1}^{n} x_i}{n} \tag{9}$$

For the ungrouped data of Table 8.1, $n = 100$, and the sum of the scores ($\sum x_i$) equals 4262. Therefore

$$\bar{x} = \frac{4262}{100} = 42.62$$

If the mean of the scores of another group of 100 students taking the same test were 38.24, we might reasonably state that the performance of the first group of students was better than that of the second, although nothing could be said of the performance of individual students in either group.

The Arithmetic Mean of a Frequency Distribution

To determine the arithmetic mean of a frequency distribution, we assume that the variates are uniformly distributed throughout each class. With this assumption, the class mark is representative of the class just as the arithmetic mean is representative of the entire distribution. In fact, for a uniform distribution the class mark *is* the arithmetic mean of the variates in the class. Then if k is the number of classes, f_i is the frequency of the ith class, and x_i is the class mark,

$$\bar{x} = \frac{x_1 f_1 + x_2 f_2 + \cdots + x_k f_k}{f_1 + f_2 + \cdots f_k} \tag{10}$$

Formula 10 may be also written

$$\bar{x} = \frac{\sum_{i=1}^{k} x_i f_i}{\sum_{i=1}^{k} f_i} \tag{11}$$

Obviously $f_1 + f_2 + \cdots + f_k = \sum_{i=1}^{k} f_i = n$, the total number of variates.

Table 8.4 Scores Made by Students in a Mathematics Test

Class Interval	Mid-Value: x_i	Frequency: f_i	Product: $x_i f_i$
55.5–60.5	58	9	522
50.5–55.5	53	11	583
45.5–50.5	48	15	720
40.5–45.5	43	28	1204
35.5–40.5	38	16	608
30.5–35.5	33	10	330
25.5–30.5	28	8	224
20.5–25.5	23	3	69
		$\sum f_i = 100$	$\sum x_i f_i = 4260$

To calculate the mean of the 100 scores on the mathematics test, we add a column for the products $x_i f_i$, as in Table 8.4, and from Equation 11 we have

$$\bar{x} = \frac{4260}{100} = 42.60$$

It is worthy of note that the mean calculated from the frequency distribution, 42.60, is remarkably close to the mean of the undistributed data, 42.62, which is, of course, the exact mean of the data. This is an indication that the selection of class limits in the frequency distribution was a good choice.

In Figure 8.5, the frequency polygon of the distribution, we have indicated the mean on the scale of variates. We see that it lies at or near the center of the data, which is the reason for the term "measure of central tendency" in describing the mean.

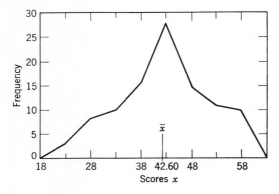

Figure 8.5

The Median of Ungrouped Data

The *median* is defined as the middle variate of a set of variates arranged in order of magnitude. Unlike the mean, the median cannot be found from an unordered set of variates, such as the original set of test marks in Section 42, since it is impossible to pick the middle variate from such an array. When the number of variates is odd, the median is the middle variate; when the number of variates is even, the median is the arithmetic mean of the two middle variates. The following examples will illustrate our remarks.

EXAMPLE 1. Find the median of the set of bowling scores:

115, 112, 138, 140, 150, 130, 125, 110, 148

Solution: The scores, arranged in order, are

110, 112, 115, 125, *130*, 138, 140, 148, 150

The middle score is 130, which is the median score.

EXAMPLE 2. Find the median of the set of mathematics scores:

62, 75, 87, 91, 69, 80, 89, 93, 70, 84, 90, 95

Solution: The scores, arranged in order, are

62, 69, 70, 75, 80, *84*, *87*, 89, 90, 91, 93, 95

The two middle scores are 84, and 87. The median is

$$\frac{84 + 87}{2} = 85.5$$

When the data consist of a large number of scores, we may determine the middle variate of an odd number of them by the formula

$$\text{position of middle term} = \frac{n + 1}{2} \qquad (12)$$

where n is the number of variates. For an even number, the two middle variates are those corresponding to

$$\frac{n}{2} \quad \text{and} \quad \frac{n}{2} + 1$$

In using these formulas it is important to keep in mind that they *locate* the median and are not formulas for the *value* of the median.

EXAMPLE 3. Find the position of the median of 101 measurements.
Solution: The position of the middle term is $(101 + 1)/2 = 51$. The median is the 51st variates.

EXAMPLE 4. Find the positions of the two middle variates of 100 measurements.
Solution: The positions of the two middle variates are $100/2$ and $(100/2) + 1$. The median is the average of the 50th and 51st variates.

The Median of a Frequency Distribution

As in the calculation of the arithmetic mean of a frequency distribution, we assume that the variates are uniformly distributed throughout each interval. We do not, however, take the class work to represent the entire class, since this would defeat our purpose of choosing a particular variate for the median. To calculate the median, it is convenient to add a cumulative frequency column to the table, as in Table 8.5.

In calculating the median of a frequency distribution, we do not distinguish between an odd or even number of variates, but locate the median by counting $n/2$ terms from either end, n being the total number of variates. In Table 8.5, $n/2$ equals 50, and it is clear from the cumulative frequency column that this variate lies in the class interval 40.5–45.5, called the median *class*. Then the 13th variate above 40.5 will be the median. Now since there are a total of 28 variates in the median class, assumed to be uniformly distributed, and since the width of the median class is 5, the difference between successive variates is $5/28$. Then the 13th variate above 40.5 will equal $40.5 + 13(5/28)$ or 42.82, which is the median. If we had counted *down* from the highest score, we would have had Median $= 45.5 - 15(5/28)$ or 42.82.

Table 8.5 Scores Made by Students in a Mathematics Test

Class Interval h	Mid-Value x	Frequency f	Cumulative Frequency
55.5–60.5	58	9	100
50.5–55.5	53	11	91
45.5–50.5	48	15	80
40.5–45.5	43	28	65
35.5–40.5	38	16	37
30.5–35.5	33	10	21
25.5–30.5	28	8	11
20.5–25.5	23	3	3
Total			

Table 8.6 Grade-Point Averages of 947 College Freshmen

Class Interval	Mid-Value	Frequency	Cumulative Frequency
3.795–3.995	3.89	2	947
3.595–3.795	3.69	52	945
3.395–3.595	3.49	161	893
3.195–3.395	3.29	170	732
2.995–3.195	3.09	190	562
2.795–2.995	2.89	175	372
2.595–2.795	2.69	125	197
2.395–2.595	2.49	51	72
2.195–2.395	2.29	17	21
1.995–2.195	2.09	2	4
1.795–1.995	1.89	2	2

The process just described may be summarized by the formula

$$\text{Median} = x_l + \left(\frac{n}{2} - f_c\right)\frac{h}{f_m} \qquad (13)$$

where x_l is the lower end-value of the median class
f_c is the cumulative frequency below the median class
f_m is the frequency of the median class
h is the width of the class interval
n is the total number of variates

For the grade-point averages in Table 8.6

$$\text{Median} = 2.995 + \left(\frac{947}{2} - 372\right)\frac{0.2}{190} = 3.102$$

The Mode of Ungrouped Data

The mode of a set of ungrouped data is the variate that occurs with greatest frequency. For the set of test scores

32, 33, 35, 36, 40, 40, 41, 41, 41, 44, 45, 45, 46, 48, 51

the mode is 41, since it occurs with greatest frequency. The mode may be applied to descriptive as well as numerical data; if a survey shows that more people prefer blue for the color of soap, then the *modal color* is blue. The fact that one of the several definitions of the word *mode* to be found in the dictionary is "the prevailing fashion" is indicative of its use as a descriptive as well as a numerical measure. Although we shall discuss the advantages and disadvantages of the three measures of central tendency at the end of this section, it is important to state here that a serious disadvantage of the mode is that it may not exist or that it may not be unique. In Example 1 of this section there is no bowling score that occurs with greatest frequency, and in the set of 100 mathematics scores (ungrouped) the greatest frequency occurs three times: 44, 45, 49; each has a frequency of 7. We see that the bowling scores have no mode, and the mathematics scores have three modes.

The Mode of a Frequency Distribution

For a frequency distribution we first define the *modal class* as the class with the greatest frequency. In Table 8.2 the modal class is 40.5–45.5, since its frequency, 28, is greater than the frequency of any other class. While there are a number of ways to define the mode of a frequency distribution, we shall take the class mark of the modal class as an approximation to the mode. The mode defined in this way is sometimes called the *crude mode*, since it is an approximation. The mode of the distribution of the 100 mathematics test scores is then 43.

Characteristics of the Mean, Median, and Mode

Certain characteristics of the mean which recommend its use are (1) it is reasonably easy to compute, (2) it always exists, (3) it is unique, and (4) it uses each measure. The fact that the mean uses each measure may also be looked on as a disadvantage, since one very high or very low score may affect the mean to a great extent. Consider two sets of test marks:

Student A   54, 51, 47, 30, 44, 42   Mean: 44.57

Student B   53, 50, 46, 44, 43, 40   Mean: 46.00

If 46 is the minimum average required for a *B* in the course, Student A who has, with one exception, higher scores than Student B, fails to get the *B* grade, while Student B makes the *B*. Of course, one very high score may raise the average of a student to a higher course grade, which would be considered an advantage rather than a disadvantage.

Some desirable characteristics of the median are (1) it is not affected by extremes at either end of the distribution, (2) it is unique, (3) it always exists, and (4) for ungrouped data it requires no calculations.

The advantages of the mode are (1) it requires little or no calculation, and (2) it can be used in describing nonquantitative data, such as color preference.

As we have stated, its great disadvantages are that it may not exist, if there is no greatest frequency, and that it may not be unique if more than one class has the same highest frequency.

EXERCISE 43

Measures of Central Tendency

1. Find the mean, median, and mode of the ungrouped data of Problem 2, Exercise 42.
2. Find the mean, median, and mode of the frequency distribution of Problem 2, Exercise 42.
3. Find the mean, median, and mode of the frequency distribution of Problem 3, Exercise 42.
4. Find the mean, median, and mode of the frequency distribution of Problem 4, Exercise 42.
5. Find the mean, median, and mode of the ungrouped data of Problem 3, Exercise 42.
6. Compare each of the measures found in Problem 3 of this section with the like measures found in Problem 5.
7. Find the mean, median, and mode of the frequency distribution of Problem 5, Exercise 42.

Write out in full each of the following summations.

8. $\sum_{i=1}^{6} x_i$ 9. $\sum_{i=1}^{4} 3x_i$ 10. $\sum_{i=1}^{5} (x_i - 2)$ 11. $\sum_{i=1}^{5} 3(x_i - 3)$

Evaluate each of the following summations for $x_1 = 2$, $x_2 = 3$, $x_3 = 4$, $x_4 = 5$, and $x_5 = -2$.

12. $\sum_{i=1}^{5} x_i$ 13. $\sum_{i=1}^{5} x_i^2$ 14. $\sum_{i=1}^{5} 3(x_i - 2)$

15. The mean of a distribution of 100 variates is 35, and the mean of a second distribution of 60 variates is 46. Find the mean of the combined distribution.
16. Using M_1 and M_2 for the means of two distributions, and f_1 and f_2 for their respective frequencies, write the formula for the mean, M, of the combined distribution.
17. Under what conditions will the mean of two combined distributions be the average of their separate means?
18. Choose three small distributions to illustrate the fact that the medians of two separate distributions can not be used to find the median of their combined distribution.

19. If the average price of butter in New York is higher than the average price in Wisconsin, is it possible to purchase a pound of butter in New York for less than it is possible to purchase a pound in Wisconsin?
20. If the modal number of suits for American men is 4, does this mean that most American men have four suits? Explain your answer.
21. Would the number of railroad cars available for a suburban run from a large city be based on a modal or a mean traffic situation? Explain your answer.
22. A mathematics teacher discovered that one paper, marked 56, was incorrectly copied in an ungrouped distribution as 46. If the mean was 40.6 and there were 40 papers, determine the correct mean.
23. In a school with several sections of college algebra, all sections take the same tests. If the median of one class was consistently lower than the others on each test, would the median be most likely raised by special attention to the lower few students, the upper few students, or the "average" students? Explain your answer.

44. MEASURES OF DISPERSION

A college student looking for a comfortable place to spend his spring vacation finds that the average daily temperature at each of two resorts, for the same period the year before, was 72°. At first thought it seems that, as far as the weather is concerned, the two resorts are about the same. However, he consults the files of the local papers at each of these resorts and finds that for this period the hourly temperatures on a particular day were as follows:

Resort A: 69° 70° 71° 74° 73° 76° 78° 73° 72° 71° 70° 67°
Resort B: 40° 52° 64° 74° 82° 95° 97° 98° 84° 78° 58° 42°

Clearly the arithmetic mean alone is of little value in comparing the temperatures of the two resorts. Resort B, with the same average daily temperature as Resort A, would be far less desirable if marked variations in temperature are to be avoided. We could easily construct data to show that the same situation may occur when the median or the mode is used as the only descriptive measure. What these measures fail to indicate is the *dispersion* of the data, that is, the spread of the data on either side of the mean. We shall indicate several methods of measuring the dispersion of the data and discuss their relative merits.

The Range

One very simple measure of dispersion is the *range*. For ungrouped data the range is the difference between the largest and smallest values of the data.

For a frequency distribution, the range is the difference between the highest and lowest values of the class boundaries. For Resort A the range is 78°–67°, or 11°. For Resort B, the range is 98°–40°, or 58°. For the 100 test scores in Table 8.2, the range is 55.5–20.5, or 35. An obvious defect of the range as a measure of dispersion is that it takes into account only the largest and smallest measures; if all the temperatures of Resort B were 72° except the highest, 98°, and the lowest, 40°, the range would remain 58°, but this would represent a different situation than before.

The Mean Deviation of Ungrouped Data

Since we can indicate the dispersion of a set of data by the differences between the variates and their mean, the *mean deviation* (that is, the average deviation from the mean) is a second measure of dispersion that we may consider. Before giving this measure, we shall state two tentative definitions.

Definition 5. The *deviation* of a variate from the mean is the difference between the variate and the mean.

If d is the deviation of any variate from the mean, then

$$d_i = x_i - \bar{x} \qquad (14)$$

Definition 6. The *mean deviation* of a set of variates is the sum of the deviations divided by the number of variates, that is, the arithmetic mean of the deviations.

If $\bar{d}$ is the mean deviation, then

$$\bar{d} = \frac{\sum_{i=1}^{n}(x_i - \bar{x})}{n} \qquad (15)$$

To determine the mean deviation of the temperatures in the two resorts, we construct Table 8.7 with the appropriate columns. For convenience, the temperatures have been written in the order of magnitude.

The fact that the sum of the deviations for each of these two widely different distributions equals zero suggests that this may be true for all distributions. We shall show at the end of this section that this is indeed the case; we shall prove that for every distribution the sum of the deviations, where Definition 6 is implied, equals zero. This being the case, the mean deviation as defined by Definition 6 is useless as a measure of dispersion, since all distributions have mean deviations of zero. This defect, however, is easily remedied if we consider all deviations as positive differences, that is, if we use the absolute values of the deviations.

Definition 7. The *deviation* of a variate from the mean is the absolute value of the difference between the variate and the mean.

Table 8.7 Calculation of Mean Deviation

Resort A		Resort B	
x_i	$d = x_i - 72$	x_i	$d = x_i - 72$
78	6	98	26
76	4	97	25
74	2	95	23
73	1	84	12
73	1	82	10
72	0	78	6
71	−1	74	2
71	−1	64	−8
70	−2	58	−14
70	−2	52	−20
69	−3	42	−30
67	−5	40	−32
	$\sum d_i = 0$		$\sum d_i = 0$

If d is the deviation of any variate from the mean, then

$$d_i = |x_i - \bar{x}| \tag{14a}$$

Definition 8. The *mean deviation* of a set of variates is the sum of the abolute values of the deviations divided by the number of variates, that is, the arithmetic mean of the absolute values.

If $\bar{d}$ is the mean deviation, then

$$\bar{d} = \frac{\sum_{i=1}^{n} |x_i - \bar{x}|}{n} \tag{15a}$$

By taking the absolute value of the deviations in Table 8.7, it is easily verified that for Resort A, $\bar{d} = 28/12 = 2.33$, and that for Resort B, $\bar{d} = 208/12 = 17.33$. These values of the mean deviation now clearly indicate the difference in the two sets of data as far as the dispersion of the data is concerned. If a comfortable "spot" is a place where the temperature is mild and fairly constant, Resort A is the best choice.

The Mean Deviation of a Frequency Distribution

In determining the mean deviation of a frequency distribution, the deviations of all variates in a given class are measured from the class mark. The mean deviation of the distribution, $\bar{d}$, is then given by the formula

$$\bar{d} = \frac{\sum_{i=1}^{k} f_i |x_i - \bar{x}|}{\sum_{i=1}^{k} f_i} \tag{16}$$

where k is the number of classes, as in Equation 10 for calculating the mean.

The calculation of the mean deviation is facilitated by the columns $|x_i - \bar{x}|$ and $f_i |x_i - \bar{x}|$ in the frequency distribution, as in Table 8.8.

Table 8.8 Calculation of Mean Deviation of a Frequency Distribution

x_i	f_i	$x_i f_i$	$|x_i - 42.6|$	$f_i |x_i - 42.6|$
58	9	522	15.4	138.6
53	11	583	10.4	114.4
48	15	720	5.4	81.0
43	28	1204	0.4	11.2
38	16	608	4.6	73.6
33	10	330	9.6	96.0
28	8	224	14.6	116.8
23	3	69	19.6	58.8
Σ	100	4260		690.4

From Formula 16 we have

$$\bar{d} = \frac{690.4}{100} = 6.90$$

The mean deviation, unlike the range, includes all of the variates. It is fairly easy to compute, particularly if a calculator or even a slide rule is available.

The Standard Deviation of Ungrouped Data

The standard deviation is a measure of dispersion which is very widely used for reasons that will be discussed in Section 45 in connection with the normal curve. It converts all of the deviations to positive numbers by taking their squares, and is defined as follows, using the symbol σ, read "sigma," for the standard deviation:

Definition 9:

$$\sigma = \sqrt{\frac{\sum_{i=1}^{n} (x_i - \bar{x})^2}{n}} \tag{17}$$

Table 8.9 Calculation of Standard Deviation

	Resort A			Resort B	
x_i	$x_i - 72$	$(x_i - 72)^2$	x_i	$x_i - 72$	$(x_i - 72)^2$
78	6	36	98	26	676
76	4	16	97	25	625
74	2	4	95	23	529
73	1	1	84	12	144
73	1	1	82	10	100
72	0	0	78	6	36
71	−1	1	74	2	4
71	−1	1	64	−8	64
70	−2	4	58	−14	196
70	−2	4	52	−20	400
69	−3	9	42	−30	900
67	−5	25	40	−32	1024
		$\sum d^2 = 102$			$\sum d^2 = 4698$
		$\sigma = \sqrt{\dfrac{102}{12}} = 2.92$			$\sigma = \sqrt{\dfrac{4698}{12}} = 19.78$

For the two resorts previously discussed, we construct Table 8.9. The values of σ do not differ greatly from the values of the mean deviations, which we found were 2.33 and 17.33, respectively.

The Standard Deviation of a Frequency Distribution

In a manner analogous to the defining of the mean deviation of a frequency distribution, we define the standard deviation, σ.

Definition 10:

$$\sigma = \sqrt{\dfrac{\sum\limits_{i=1}^{k} f_i(x_i - \bar{x})^2}{\sum\limits_{i=1}^{k} f_i}} \tag{18}$$

For the 100 mathematics scores, we construct Table 8.10 with the columns $(x_i - \bar{x})^2$ and $f_i(x_i - \bar{x})^2$ for the particular value of $\bar{x} = 42.6$

$$\sigma = \sqrt{\dfrac{7884}{100}} = 8.88$$

The particular significance of the standard deviation will be discussed in Section 45.

Table 8.10 Calculation of Standard Deviation of a Frequency Distribution

x_i	f_i	$x_i - 42.6$	$(x_i - 42.6)^2$	$f_i(x_i - 42.6)^2$
58	9	15.4	237.16	2134.44
53	11	10.4	108.16	1189.76
48	15	5.4	29.16	437.40
43	28	0.4	0.16	4.48
38	16	−4.6	21.16	338.56
33	10	−9.6	92.16	921.60
28	8	−14.6	213.16	1705.28
23	3	−19.6	384.16	1152.48
Σ	100			7884.00

Proof that the Sum of the Deviation Equals Zero

By definition

$$\sum_{i=1}^{n} d_i = \sum_{i=1}^{n} (x_i - \bar{x}) \tag{1}$$

Using Equation 4, Section 43, we have

$$\sum_{i=1}^{n} d_i = \sum_{i=1}^{n} x_i - \sum_{i=1}^{n} \bar{x} \tag{2}$$

Since $\bar{x}$ is a constant, by Equation 5, Section 43

$$\sum_{i=1}^{n} \bar{x} = n\bar{x} \tag{3}$$

Substituting Equation 3 in Equation 2,

$$\sum_{i=1}^{n} d_i = \left\{\sum_{i=1}^{n} x_i\right\} - n\bar{x} \tag{4}$$

By definition,

$$\bar{x} = \frac{\sum_{i=1}^{n} x_i}{n} \quad \text{so that} \quad n\bar{x} = \sum_{i=1}^{n} x_i \tag{5}$$

Substituting Equation 5 in Equation 4, we have

$$\sum_{i=1}^{n} d_i = \sum_{i=1}^{n} x_i - \sum_{i=1}^{n} x_i = 0$$

which proves that the sum of the deviations equals zero.

EXERCISE 44

MEASURES OF DISPERSION

1. Find the mean deviation and the standard deviation of the ungrouped data of Problem 2, Exercise 42.
2. Find the mean deviation and the standard deviation of the ungrouped data of Problem 3, Exercise 42.
3. Find the mean deviation and the standard deviation of the following frequency distribution of Arithmetic Scores.

Arithmetic Scores

Class	x_i	f_i
0.5–5.5	3	8
5.5–10.5		16
10.5–15.5		56
15.5–20.5		68
20.5–25.5		96
25.5–30.5		100
30.5–35.5		92
35.5–40.5		80
40.5–45.5		72
45.5–50.5		54
50.5–55.5		44
55.5–60.5		36
60.5–65.5		18
65.5–70.5		14

4. Find the mean deviation and the standard deviation of the following frequency distribution of Reading Comprehension Scores.

Reading Comprehension Scores

Class	x_i	f_i
1.5–4.5	3	2
4.5–7.5	6	4
7.5–10.5		16
10.5–13.5		30
13.5–16.5		46
16.5–19.5		40
19.5–22.5		32
22.5–25.5		17
25.5–28.5		5
28.5–31.5		6
31.5–34.5		1

5. Find the standard deviation of the following frequency distribution of the I.Q.'s of college freshmen.

I.Q. of Freshmen

Class	x_i	f_i	$x_i f_i$
69.5–79.5	74.5	2	
79.5–89.5		18	
89.5–99.5		75	
99.5–109.5		69	
109.5–119.5		20	
119.5–129.5		14	
129.5–139.5		2	

45. FREQUENCY DISTRIBUTIONS AND THE NORMAL CURVE

In Figure 8.3, Section 42, the frequency polygon was drawn by connecting the points, which represented the frequencies of specific scores on the test, by straight lines. Variates of this type are called *discrete* variates, since they assume only certain values within the range. In the distribution of test scores, these values were positive integers between 21 and 60. In distributions such as the heights of men or the temperatures during a given period, the variates can assume all of the values within the given range and are, therefore, called *continuous* variates. While two consecutive recorded temperatures may be 68° and 69°, it is clear that the temperature, in changing from 68° to 69°, must pass through all intermediate values. In distributions where the data are large, we may assume the variates to be continuous and draw a *frequency curve* in place of a frequency polygon. Figure 8.6 shows a frequency curve for the temperatures during a certain 12-hour period in Chicago.

The Normal Distribution

Frequency curves may assume a great variety of forms, depending on the data which they represent. Of particular interest in theoretical and applied statistics is the *normal distribution* and the related *normal curve*. In the 18th century mathematicians observed that the errors associated with a large number of careful measurements[3] formed a frequency distribution for which

---

[3] By the term "careful measurements" we mean that the errors are not *mistakes* but the result of the inherent imperfections of measuring instruments and the vision of the observer.

45. Frequency Distributions and the Normal Curve   269

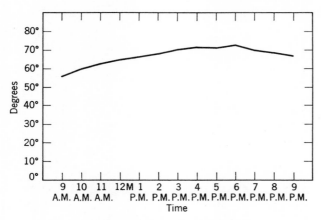

Figure 8.6 Hourly temperatures.

the frequency curve was a bell-shaped symmetrical curve, as in Figure 8.7. As in the graph of the quadratic function, the normal curve has an equation, but we shall not state it here. The importance of the normal distribution is that it is typical of many kinds of measurements, such as physical and mental measurements of men, the life span of electric light bulbs, or the grades of a large number of students taking a college-entrance test.

Properties of a Normal Distribution

We shall state without proof certain properties of a normal distribution which may be derived from its equation in the same manner that properties of the parabola, such as the axis of symmetry, may be derived from the function $y = ax^2 + bx + c$.

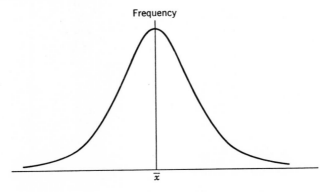

Figure 8.7

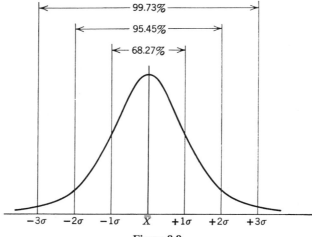

Figure 8.8

Property 1. The mean, median, and mode of a normal distribution are equal to each other.

Property 2. The vertical line through the point representing the equal measures of central tendency is the axis of symmetry of the curve.

Property 3. Approximately 68% of the variates lie within one standard deviation of the mean, and approximately 95% of the variates lie within two

Table 8.11 Scores Made by Students on a Mathematics Test

Scores	Frequency	Product
55	2	110
50	4	200
45	6	270
40	9	360
35	6	210
30	1	30
25	2	50
10	1	10
	31	1240

Mean $= \dfrac{1240}{31} = 40$

Mode $= 40$

Median $= 40$

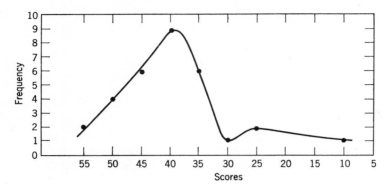

Figure 8.9 Scores made by students on a mathematics test.

standard deviations of the mean. Almost 100% of the variates lie within three standard deviations of the mean.

Figure 8.8 illustrates these properties of the normal curve. The units on the horizontal scale are in terms of standard deviations.

Property 1 may be used to *suggest* that a distribution may be normal if the three measures of central tendency are equal, or nearly so. However, the distribution of Table 8.11, in which the mean, median, and mode are all equal to 40, is not a normal distribution, as the frequency curve in Figure 8.9 clearly indicates.

Property 2 may also be used to suggest that a distribution is normal, if symmetrical about the vertical axis. However, the frequency curve in Figure 8.10, while perfectly symmetrical about the vertical line through the median score, represents a bimodal distribution (two modes) and is, therefore, not a normal distribution.

Property 3 is helpful in making certain predictions and is the reason for the importance of the standard deviation as a measure of dispersion. Suppose we

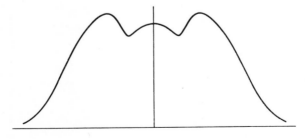

Figure 8.10

know that the average daily temperatures at a certain resort are normally distributed, with a mean of 72° and a standard deviation of 12°. We can then estimate that approximately 68% of the days will have temperatures between 60° and 84°. Whether or not this is indicative of a satisfactory place to spend a vacation is, of course, a matter of personal preference.

Misuses of the Normal Curve

The mathematical derivation of the equation of the normal curve assumes that the distribution is composed of an infinite number of variates, although, practically, this means a very large number of variates. Therefore caution

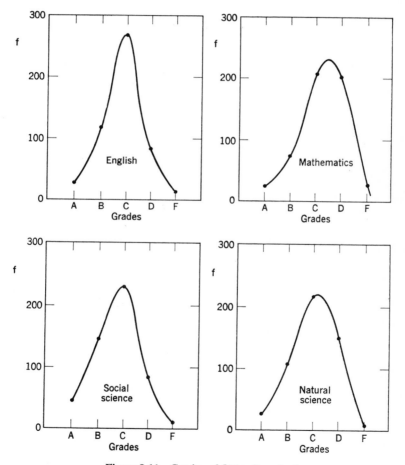

Figure 8.11 Grades of 516 college freshmen.

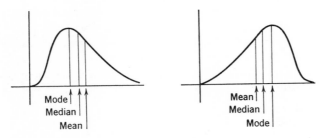

Figure 8.12

must be used when applying the properties of a normal distribution to a relatively small number of variates. An instructor who grades a class of 30 students on the "curve" would be restricted to about two A's and would have to give about two F's. Obviously a class of 30 students might easily be worthy of no A's, or might have as many as five or six. It is also possible that in a class of 30 students none would do so poorly that they would fail the course. On the other hand, distributions involving a large number of variates are not necessarily normal distributions. Figure 8.11, which shows the grades of 500 college freshmen in four subjects indicates that the distribution is only approximately normal, although the number of variates is fairly large.

Frequency curves such as those in Figure 8.12, which do not possess the symmetry of the normal curve, are said to be *skewed*. In these distributions the mean, median, and mode tend to occupy the relative positions indicated. In such cases the median is usually a more reliable measure of the central tendency of the data. This is because the mean has been affected by extreme values and does not accurately describe the data. Thus, if the salaries of five office workers are $100, $100, $110, $115, and $500, the mode is $100, the median is $110, and the mean is $185.

EXERCISE 45

Frequency Distribution and the Normal Curve

In the following problems use 68% and 95% for the number of cases indicated in Figure 8.8.

1. The distribution of I.Q.'s as measured by a certain test is normal, with a mean of 100 and a standard deviation of 15. Find, for 5000 I.Q.'s:

    (a) The number of I.Q.'s above 115.
    (b) The number of I.Q.'s above 130.
    (c) The number of I.Q.'s below 85.
    (d) The number of I.Q.'s below 70.

2. Discuss the answers to Problem 1 if the 5000 cases considered are the I.Q.'s of college freshman.
3. Discuss the answers to Problem 1 if the 5000 cases are winners of a National Merit Scholarship.
4. Arrange the ungrouped scores on the mathematics test discussed in Section 42 in a table in order, from highest to lowest.
   (a) Find the mean.
   (b) Find the median.
   (c) Find the mode.
   (d) Find the standard deviation, using a table as follows:

x	f	$x - \bar{x}$	$(x - \bar{x})^2$	$f(x - \bar{x})^2$
21	2	21.62	467.42	934.84

   (e) Calculate the number of scores one standard deviation and two standard deviations above and below the mean, using 68% and 95%.
   (f) Compare the number of cases found in e with the number of cases found by actual count.
   (g) In what ways does the distribution resemble a normal distribution?
   (h) In what ways does the distribution differ from a normal distribution?
5. Answer the questions in Problem 4 for the ungrouped data of Problem 5, Exercise 42.

9

What is Modern Mathematics?

46. CHARACTERISTICS OF MODERN MATHEMATICS

Modern mathematics extends and, in many respects, summarizes the rich content of traditional mathematics. There are, however, certain characteristics which distinguish modern mathematics from much of what preceded it. Some of the most important of these characteristics are:

1. *An evolutionary dominance of the set-theoretic point of view.* Some mathematicians believe that the theory of sets may be the single most important development in mathematics in the last hundred years. It is a concept that has been used to advantage to clarify, unify, enrich, and extend many domains of mathematics and to influence profoundly the study of the foundations of mathematics.

2. *An explosive increase in the range and application of the subject matter of mathematics.* It has been estimated that mathematical knowledge has been doubling every 15 or 20 years since the 1800's and that more original mathematics has been produced in the past half century than in all previous history. Much of this knowledge has been brought to bear on the solution of problems in such new fields as atomic energy, space science, computer development and application, and the use of game theory in business and economics.

3. *A fundamental trend toward higher levels of abstraction.* Mathematicians have always been partial to abstraction. They conceive geometries that explore the effects of distortion and algebras that are devoid of numbers. They concern themselves with a space of infinite dimensions and with numbers having differing orders of infinity. They freely prefer to explore new abstractions rather than develop new applications. Paradoxically, such excursions into pure theory have tended to create new branches of useful mathematics while uncovering many of the profound interrelations among the fundamental branches of mathematics.

4. *An increased concern with the structure of mathematics.* One result of the far-ranging attempts to unify mathematics has been a search for a single, consistent set of axioms on which all of mathematics could be based. In 1931, Kurt Gödel proved this to be impossible in one of the most influential theorems of modern mathematics, namely, that no useful branch of mathematics can be constructed on a consistent set of axioms without raising questions which cannot be answered within the structure defined by the axioms themselves. The possibility that this theorem may lead to entirely new theories of mathematics as a whole has placed a greater emphasis on the analysis of logical structure than was present in traditional mathematics.

In this chapter we will introduce several topics which have a place in modern mathematics. Since, for the most part, contemporary developments are too abstract and esoteric for our purposes, we will necessarily restrict our considerations to those rather elementary aspects of several areas which reflect the nature and flavor of modern mathematics.

47. BOOLEAN ALGEBRA

Boolean algebra is named after George Boole (1815–1864), an English logician and mathematician, who was the first to design an algebraic format for the study of logic. In recent years, the abstract symbolic logic of Boole has been applied to studies concerned with the foundations of mathematics and to such applications as the analysis of electrical networks, the design of computer circuits, and the solution of problems in the social and natural sciences.

Ordinary algebra, as noted in Section 17, is based on the field axioms as applied to the real number system. Boolean algebra, on the other hand, is based on a different set of axioms and a different number system. Actually, few branches of mathematics have received a greater diversity of axiomatic treatments than has Boolean algebra, so that today there exist a wide variety of approaches to the study. The approach we have chosen is one of the most common treatments.

The following terms will be accepted as undefined: set, element, and belongs to. We now define the system as follows.

A Boolean algebra is a set of elements $B = \{a, b, c, \ldots\}$, with two binary operations $(+)$ and $(\cdot)$, called addition and multiplication, respectively, and an equivalence relation, $=$, satisfying the following axioms.[1]

1. *Commutative axiom.* For all elements a and b in the set (that is, belonging to the set),

$$a + b = b + a \quad \text{and} \quad a \cdot b = b \cdot a$$

2. *Identity axiom.* There exist elements 0 and 1 such that for each element a in the set,

$$a + 0 = a \quad \text{and} \quad a \cdot 1 = a$$

3. *Distributive axiom.* For all a, b, c in the set,

$$a + (b \cdot c) = (a + b) \cdot (a + c)$$
and
$$a \cdot (b + c) = (a \cdot b) + (a \cdot c)$$

4. *Complementation axiom.* For each element a in the set, there exists an element a' in the set such that

$$a + a' = 1 \quad \text{and} \quad a \cdot a' = 0$$

Certain similarities as well as differences between this set of axioms and the field axioms of ordinary algebra are immediately noticeable. The commutative axiom and the identity axiom are certainly analogous to the corresponding field axioms for addition and multiplication. However, the distributive axiom is different because it applies to both operations whereas in the field axioms only one operation is distributive with respect to the other. The complementation axiom also differs from that of the field axioms in an obvious and significant way. For example, $a + a' = 0$ in the field axioms, but $a + a' = 1$ in the Boolean axioms.

It is apparent that a perfect symmetry exists between the two operations in the set of Boolean axioms. By interchanging the operations $(+)$ and $(\cdot)$ and the elements 0 and 1 throughout the set of axioms, we could generate the identical set of axioms. This property is called *duality*. The principle of duality is an important attribute of Boolean algebra, and it enables us to deduce a dual statement from each statement or theorem proved in the system.

---

[1] A more generalized notation usually employs the symbols $\cup$ and $\cap$ for the two operations and the symbols z and u for the respective identity elements. The notation used here is chosen because it is simple, yet adequate for our purposes.

278   What is Modern Mathematics?

We now consider a specific system which presents the structure of Boolean algebra in a simplistic, yet useful, form. Instead of the general elements $a, b, c, \ldots$, we will limit our set to just the two elements 0 and 1. The relations between the two elements can be displayed by means of the following tables:

Operation (+)			Operation (·)			Complementation
	0	1		0	1	$0' = 1$
0	0	1	0	0	0	$1' = 0$
1	1	1	1	0	1	

The property of duality is strikingly displayed in the following listing which places the results of the operations (+) and (·) side by side:

$$0 + 0 = 0 \qquad 1 \cdot 1 = 1$$
$$1 + 0 = 1 \qquad 0 \cdot 1 = 0$$
$$0 + 1 = 1 \qquad 1 \cdot 0 = 0$$
$$1 + 1 = 1 \qquad 0 \cdot 0 = 0$$

In each case if, in the first column, we replace 0 by 1, 1 by 0, and (+) by (·), we obtain the corresponding relation in the second column. Note that although, in general, the elements 0 and 1 and the operations (+) and (·) behave much like their counterparts in ordinary algebra, there is an important difference expressed in the relation $1 + 1 = 1$. Therefore, the 0 and 1 of Boolean algebra are not to be confused with the 0 and 1 of either the real number system wherein $1 + 1 = 2$ or the binary number system wherein $1 + 1 = 10$.

The relation $1 + 1 = 1$ in Boolean algebra is a direct result of a general theorem that, for every element x in the system, $x + x = x$. This theorem is derived directly from the four axioms which define the system. We will sketch the proof as an example of deductive reasoning in an abstract system.

Theorem 1. For each element x in the set B, $x + x = x$.

Proof:

$x + 0 = x$	By Axiom 2—identity axiom for addition
$x + x \cdot x' = x$	By Axiom 4—complementation axiom for multiplication
$(x + x) \cdot (x + x') = x$	by Axiom 3—distributive axiom
$(x + x) \cdot 1 = x$	by Axiom 4—complementation axiom for addition
$x + x = x$	by Axiom 2—identity axiom for multiplication

As another example of the proof of a theorem in Boolean algebra, consider the expression $x + 1$, where x is any element of a given Boolean set. In our simplified system it is apparent that, whether $x = 0$ or $x = 1$, in each case $x + 1 = 1$. Is this true for any replacements of x in any Boolean system? The following general theorem assures us that it is.

Theorem 2. For every element x in set B, $x + 1 = 1$.
Proof:

$x + x' = 1$	by Axiom 4—complementation axiom for addition
$x + (x' \cdot 1) = 1$	by Axiom 2—identity axiom for multiplication
$(x + x') \cdot (x + 1) = 1$	by Axiom 3—distributive axiom
$1 \cdot (x + 1) = 1$	by Axiom 4—complementation axiom for addition
$(x + 1) \cdot 1 = 1$	by Axiom 1—commutative axiom for multiplication
$x + 1 = 1$	by Axiom 2—identity axiom for multiplication

As in ordinary algebra, previously proved theorems, as well as axioms, can be used to derive additional relations in the system. Consider, for example, the expression $x + (x \cdot y)$. If x and y are restricted to the values 0 and 1, it is possible to tabulate all possible values of the given expression, as follows:

x	y	$x \cdot y$	$x + (x \cdot y)$
0	0	0	0
0	1	0	0
1	0	0	1
1	1	1	1

An interesting result is apparent, namely that $x + (x \cdot y) = x$. That this is always true is proved in the following theorem.

Theorem 3. For all values of x and y in B, $x + (x \cdot y) = x$.
Proof:

$x \cdot 1 = x$	by Axiom 2—identity axiom for multiplication
$x \cdot (y + 1) = x$	by Theorem 2
$(x \cdot y) + (x \cdot 1) = x$	by Axiom 3—distributive axiom
$(x \cdot y) + x = x$	by Axiom 2—identity axiom for multiplication
$x + (x \cdot y) = x$	by Axiom 1—commutative axiom for addition

It is evident that many other theorems can be developed from the basic axioms of a Boolean system, and some additional examples will be considered

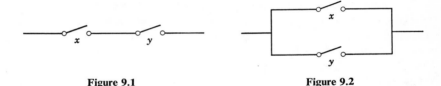

Figure 9.1   Figure 9.2

in the exercise which follow this section. The theorems and their proofs illustrate an important characteristic of modern mathematics—the logical investigation of abstract systems. We now consider one application which assigns special meaning to the elements, operations, and statements of Boolean algebra.

Switching Circuits

One of the uses of a two-element Boolean algebra is to simplify the design of electrical networks. In Figures 9.1 and 9.2, x and y represent electrical switches.

If the switches are connected into the circuit as in Figure 9.1, they are said to be in *series*, and if they are connected as in Figure 9.2, they are said to be in *parallel*. It is obvious that current will flow in the series circuit only if both x and y are closed but that current will flow in the parallel circuit if either x or y is closed or if both x and y are closed. We now assign the following interpretations to the electrical system.

Electrical network	Boolean symbol
Switch open	0
Switch closed	1
Switches connected in series	$(\cdot)$
Switches connected in parallel	$(+)$
Equivalent networks	$=$

This symbolism can be applied to the four possible situations of a series connection as shown in Figure 9.3. If the right member of the Boolean relation

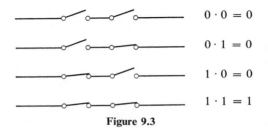

$0 \cdot 0 = 0$

$0 \cdot 1 = 0$

$1 \cdot 0 = 0$

$1 \cdot 1 = 1$

Figure 9.3

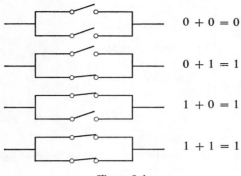

$$0 + 0 = 0$$
$$0 + 1 = 1$$
$$1 + 0 = 1$$
$$1 + 1 = 1$$

Figure 9.4

is 0, it means that the circuit is open and current will not flow. We note again that only in the last case, when both switches are closed, will the current flow. This is indicated by the element 1 in the right member of the corresponding equation.

We can analyze in the same manner, the four possible combinations of closed and open switches for a parallel connection, as illustrated in Figure 9.4.

The Boolean equations indicate that no current is flowing in the first case (since the right member is 0) but that current does flow in each of the other three cases (since the right member is 1). There is, therefore, a correspondence between these switching arrangements and the Boolean relations.

Consider now the electrical circuit illustrated in Figure 9.5. The lower part of the circuit indicates that the two switches, x and y, are connected in series. This branch can be denoted by $x \cdot y$, and since it is connected in parallel with another switch x (upper branch), the entire circuit can be symbolized by the expression

$$x + (x \cdot y)$$

But it will be recalled that, by Theorem 3, $x + (x \cdot y) = x$. This means that the entire arrangement can be replaced by a single switch x, as in Figure 9.6. In effect, as far as current flowing or not flowing is concerned, the single switch in Figure 9.6 will perform as well as the combination of three switches in Figure 9.5.

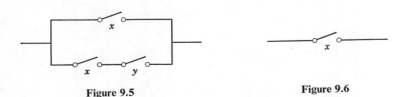

Figure 9.5 **Figure 9.6**

EXERCISE 47

BOOLEAN ALGEBRA

Write the dual of each of the following Boolean relations.

1. $x + x = x$
2. $x + 1 = 1$
3. $x + (x \cdot y) = x$
4. $a + (b + c) = (a + b) + c$
5. $(a \cdot b)' = a' + b'$
6. $x + x' \cdot y = x + y$

Complete the following tables for a two-valued Boolean system, and thereby show that in every case:

7. $(a \cdot b)' = a' + b'$

a	b	$a \cdot b$	$(a \cdot b)'$	a'	b'	$a' + b'$
0	0					
0	1					
1	0					
1	1					

8. $a \cdot (a' + b) = a \cdot b$

a	b	a'	$a' + b$	$a \cdot (a' + b)$	$a \cdot b$
0	0				
0	1				
1	0				
1	1				

9. $a' + a \cdot b = a' + b$

a	b	a'	$a \cdot b$	$a' + a \cdot b$	$a' + b$
0	0				
0	1				
1	0				
1	1				

10. $(a + b)' = a' \cdot b'$

a	b	$a + b$	$(a + b)'$	a'	b'	$a' \cdot b'$
0	0					
0	1					
1	0					
1	1					

The set $\{0, 1, x, y\}$, with $(+)$ and $(\cdot)$ defined by the following table, is a Boolean algebra. Problems 11–18 refer to this set.

(+)	0	1	x	y
0	0	1	x	y
1	1	1	1	1
x	x	1	x	1
y	y	1	1	y

(·)	0	1	x	y
0	0	0	0	0
1	0	1	x	y
x	0	x	x	0
y	0	y	0	y

Cite three examples of each of the following relations.

11. $a + b = b + a$ **12.** $a \cdot 1 = a$
13. $a + (b \cdot c) = (a + b) \cdot (a + c)$ **14.** $a + a' = 1$
15. $a \cdot a' = 0$ **16.** $a + a = a$
17. $a + 1 = 1$ **18.** $a \cdot 0 = 0$

Prove, by using the Boolean axioms, that the following relations hold true.

19. $a' \cdot (a + b) = a' \cdot b$ **20.** $a + a' \cdot b = a + b$
21. Describe each of the following switching arrangements algebraically, and indicate its simplest equivalent.

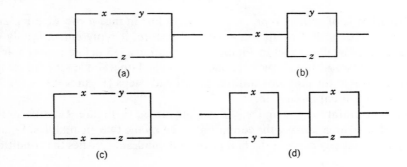

48. LINEAR PROGRAMMING

An important segment of modern mathematics is concerned with concepts and techniques for making decisions. One of the techniques used is *linear programming*—a mathematical procedure for picking the best course of action from among an infinite number of possible courses. It has been applied to the solution of certain logistics problems for military operations, to the analysis of business and economic operations, and to the allocation of men and machines in setting up production schedules.

284   What is Modern Mathematics?

Regardless of the particular area of application, problems that lend themselves to the linear programming technique always contain the following elements.

 1. There is a variable to be optimized as a *maximum* profit, or a *minimum* time, or a *least* cost, or a *greatest* yield, and so on.
 2. The variable is expressible as a linear function.
 3. There are restrictions on the domain of the variable, called *constraints*, which are expressible as linear inequalities.

The mathematical concepts behind linear programming can be illustrated by the following example. Consider a variable h which is expressible as a linear function by the relation

$$h = 2x + y \qquad (1)$$

and whose domain is defined by the following constraints:

$$x \geq 0 \qquad (2)$$
$$x \leq 8 \qquad (3)$$
$$y \geq 0 \qquad (4)$$
$$x + y \leq 10 \qquad (5)$$

Equation 1, of course, represents a straight line in the xy coordinate plane for each value of h. For all values of h, therefore, it represents a family of parallel lines as illustrated in Figure 9.7. Equations 2–5 define a closed area of the xy plane. Although they are usually graphed on a single set of axes, we have illustrated the graphing in three progressive steps (Figures 9.8–9.10) for purposes of initial discussion.

The constraints $x \geq 0$ and $x \leq 8$ are illustrated in Figure 9.8. Any point in the shaded area satisfies the conditions given by the two inequalities. Figure 9.9 shows that any point in the first or second quadrant satisfies the condition

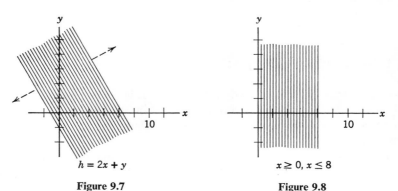

$h = 2x + y$

Figure 9.7

$x \geq 0, x \leq 8$

Figure 9.8

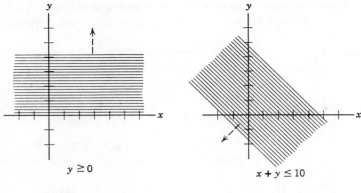

Figure 9.9   **Figure 9.10**

that $y \geq 0$. The final constraint, given by relation 5, is graphed by plotting the line $x + y = 10$ and then determining the open half plane which contains all points whose coordinates satisfy the condition $x + y \leq 10$, as illustrated in Figure 9.10.

If all four constraints are graphed simultaneously, they form a convex polygon, as shown in Figure 9.11. The vertices of the polygon, which can be determined by solving the corresponding pairs of linear equations, are $A(0, 0)$, $B(8, 0)$, $C(8, 2)$, and $D(0, 10)$. Intuitively, it is apparent that, as the family of lines $h = 2x + y$ pass through this convex polygon, the maximum or minimum value of h will always occur at a vertex, that is, at the first or last point at which the family of lines intersects the polygon. In the case of the relation $h = 2x + y$, if we substitute the values of x and y at each vertex A, B, C, and D, the corresponding values of h are 0, 16, 18, and 10, respectively.

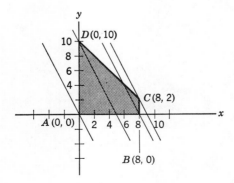

Figure 9.11

Therefore, the minimum value of h is 0, which occurs when the line $h = 2x + y$ passes through the vertex A, and the maximum value of h is 18, which occurs when the line $h = 2x + y$ passes through the vertex C.

The following example essentially repeats these steps while introducing meaning to the variables and their relations.

EXAMPLE 1. Suppose a firm manufactures two types of electronic sensors, a model M and a model N, which require the following time for processing:

	Dept. A (Fabricating)	Dept. B (Assembly)	Dept. C (Finishing)
Model M	20 min	30 min	12 min
Model N	30 min	15 min	12 min

None of the departments can work more that 40 hours per week, and Department C must work at least 10 hours per week. The profit on model M is $150 per unit, and the profit on model N is $100 per unit. How many of each model should be scheduled each week in order to maximize the firm's profit?

Solution: Let x be the number of units of model M and y the number of units of model N to be scheduled. The profit equation then becomes

$$P = 150x + 100y$$

Neither x nor y can be negative; therefore, an implied constraint is $x \geq 0$ and $y \geq 0$. Further constraints are prescribed by the number of hours each department will work to produce x units of model M and y units of model N.

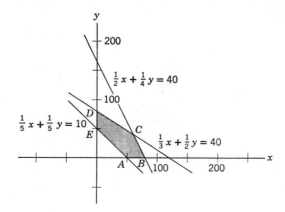

Figure 9.12

These are:

For Department A $\quad \frac{1}{3}x + \frac{1}{2}y \leq 40$

For Department B $\quad \frac{1}{2}x + \frac{1}{4}y \leq 40$

For Department C $\quad \frac{1}{5}x + \frac{1}{5}y \geq 10 \quad$ and $\quad \frac{1}{5}x + \frac{1}{5}y \leq 40$

The plotting of these conditions results in the convex polygon which is shaded in Figure 9.12. For simplicity the line $1/5\,x + 1/5\,y = 40$ is not shown, since it falls outside the polygon and, therefore, does not act as a constraint in this particular case. The coordinates of four of the five vertices of the resulting polygon are easily found since they represent x or y intercepts of the bounding lines. Thus we identify the vertices $A(50, 0)$, $B(80, 0)$, $D(0, 80)$, $E(0, 50)$. The fifth vertex C is the intersection of the two lines

$$\frac{1}{2}x + \frac{1}{4}y = 40$$
$$\frac{1}{3}x + \frac{1}{2}y = 40$$

Solving these simultaneously as a linear system, we obtain the common solution

$$x = 60, \, y = 40$$

It is now necessary to examine the condition imposed by the profit equation at each vertex. The results can be tabulated as follows:

Vertex	x	y	Profit $= 150x + 100y$
A	50	0	\$7,500
B	80	0	12,000
C	60	40	13,000
D	0	80	8,000
E	0	50	5,000

Therefore, under the given conditions, the profit will be greatest if 60 units of model M and 40 units of model N are scheduled each week.

EXERCISE 48

Linear Programming

Graph the convex polygon that corresponds to each of the following sets of constraints.

1. $x \geq 0$, $x \leq 5$, $y \geq 0$, $x + y \leq 9$
2. $x \geq 0$, $y \geq 0$, $2x + y \leq 10$, $x + 2y \leq 12$

3. $x \geq 0, y \geq 0, y \leq 6, 2x + y \geq 4, x + y \leq 7$
4. $6x + 5y \leq 30, 2x + 5y \geq 10, 3x + y \geq 6$
5. $x \geq 0, x + y \leq 6, x - 2y \geq 0$

Determine the coordinates of the vertices of the convex polygon defined by the following conditions.

6. $x \geq 0, x \leq 6, y \geq 0, x + y \leq 8$
7. $x \geq 1, y \geq 0, y \leq 6, 2x + y \leq 10$
8. $5x + 3y \geq 15, y \leq 5, x + y \leq 8, y \geq 0$

A linear programming problem is based on the following constraints: $y \geq 0$, $x + 2y \leq 14$, $3x + 2y \leq 30$, $7x + 3y \geq 21$. Find the point or points at which h is maximized in each of the following cases (Problems 9–12).

9. $h = x + y$
10. $h = 2x + y$
11. $h = x + 3y$
12. $h = x + 2y$

13. In a controlled experiment on the feeding of animals, the following two formulas of balanced food are used:

	Carbohydrates (grams per unit)	Proteins (grams per unit)	Fats (grams per unit)
Formula 1	15	20	1
Formula 2	30	10	5

It is estimated that each animal should receive daily not more than 180 grams of carbohydrates and a minimum of 120 grams of proteins and 15 grams of fats. If formula 2 costs three times as much per unit as formula 1, what combination of the two formulas will meet (or exceed) the minimum requirements at least cost?

14. In Problem 13, if the ratio of costs of formula 1 to formula 2 is 3:1, what combination of the two formulas will satisfy the minimum requirements at least cost?

49. FINITE GEOMETRY

A finite geometry is one which, by reason of the assumptions on which it is based, deals only with a finite number of elements such as points, lines, and planes. Such geometries, although simple in structure, are in every sense of the word a part of modern mathematics. In particular, they illustrate the modern viewpoints that a geometry may be nonmetrical in nature (that is, it need not be concerned with measurement, equalities, and inequalities) and that it may be based on somewhat arbitrary axioms rather than on a basis that reflects our spatial experience.

Although many different finite geometries are possible, we will select our assumptions in such a way as to confine ourselves to one of the simplest and most common systems—a plane geometry with just seven points and seven lines. We begin with the following undefined terms.

1. A set of elements $A, B, C, \ldots$ belonging to set S, called *points*.
2. Subsets of set S, called *lines*.
3. A relation of an element "belonging to" a given subset. For convenience, if a point P belongs to a subset l, we shall say that P *is on* l, or P *lies on* l. The same idea is expressed by saying that l *is on* P, or l *contains* P, or l *passes through* P.

Even though these undefined terms are suggestive of common geometric concepts, it is important to bear in mind that they refer to an abstract system and, therefore, the only properties these terms possess will be those prescribed by axioms. When we provide an interpretation which is consistent with the abstract system (as will be illustrated later in this section), then a model for the geometry will be apparent.

We specify the following statements as axioms.

1. There exists at least one line.
2. Three and only three points lie on each line.
3. Not all the points in the system lie on the same line.
4. There is one and only one line containing any two points.
5. There is at least one point common to any two lines.

Let us consider, first, an abstract illustration of the system as summarized in the following array in which the symbols $A, B, C, \ldots$ represent points and the vertical listings of three points represent lines.

Lines	1	2	3	4	5	6	7
	A	B	C	D	E	F	G
Points	B	E	F	G	A	D	F
	D	C	A	C	G	E	B

Axioms 1, 2, and 3 are satisfied by inspection. Axiom 4 requires that any two points lie on one and only one line. For example, the points A and B lie only on line 1, the points A and C lie only on line 3, and so on. (There are 21 different combinations of two points, and it can be verified that each pair lies on one and only one line.) Axiom 5 requires that every pair of lines has a common point. Thus lines 1 and 2 have point B in common, lines 1 and 3 have point A in common, lines 1 and 4 have point D in common, and so on. (Again, it is possible to verify this axiom for all 21 different pairs of lines which exist in the system.)

Interpretations of Finite Geometry

The same system can be vividly presented through a geometric representation, as in Figure 9.13. Note that one of the lines, *FDE*, is not straight, but recall that since the lines are merely subsets of points nothing compels us to use straight lines. We have drawn most of them straight only as a matter of convenience—actually all seven can be curved. In the geometric interpretation it is visually apparent that each of the five axioms is satisfied.

This geometric representation is not the only possible interpretation or model that can be constructed for the given system. For example, if the elements $A, B, C, \ldots$ represent people belonging to a certain organization and if the subsets represent committees formed from a selection of these people, then the axioms would prescribe the formation of committees as follows.

1. There must be at least one committee.
2. Each committee will consist of exactly three members.
3. Not all persons can be members of the same committee.
4. Any two persons will serve together on one and only one committee.
5. Any two committees will have at least one member in common.

It is interesting to note that under these conditions the membership of the organizations will be restricted to just seven persons, and exactly seven committees will be possible. This will be proved in Problems 4–27 which follow this section.

Other interpretations of the system can be constructed by assigning appropriate meanings to its undefined terms, as in the following.

Element	*Subclass*	*Relation of "Belonging to"*
line	plane	line in a plane
student	class	enrolled in
investment	trust fund	transaction

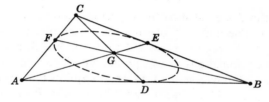

Figure 9.13

49. Finite Geometry

When different models of a system are based on the same mathematical structure but differ only in the names or ideas associated with the undefined terms of the system, we say that they are *isomorphic*. Thus, each of the interpretations we have introduced is isomorphic to the abstract system of finite geometry defined by our axioms. Isomorphism is a basic mathematical concept which ranges from such elementary examples as the correspondence between decimal and hexal numerals to highly sophisticated relations between algebraic and geometric structures.

Duality

The system of finite geometry, like the Boolean system of algebra discussed in Section 47, possesses the property of duality. Thus, by interchanging the terms "point" and "line" in the five axioms (and using the appropriate terminology for the relation "belongs to"), the following statements result.

1. There exists at least one point.
2. Three and only three lines pass through each point.
3. Not all lines in the system pass through the same point.
4. There is one and only one point common to any two lines.
5. There is at least one line containing any two points.

Although these properties appear to be easily verifiable for any given model of finite geometry, such verification, of course does not constitute a proof of the principle of duality. To illustrate such proof, as well as to furnish some interesting examples of deductive processes, we consider a few of the many theorems which can be developed from the simple base of three undefined terms and five axioms.

Theorem 1. There exists at least one point.
Proof: By Axiom 1, there exists at least one line. But, by Axiom 2, there are exactly three points on the line. Therefore, there is at least one point.

Theorem 2. No more than one point is common to any two lines.
Proof: If we assume that there are two distinct points common to each of two lines, then an immediate contradiction is established by Axiom 4. Therefore, the assumption is untenable.

Theorem 3. There is one and only one point common to any two lines.
Proof: By Axiom 5, there is at least one point common to any two lines and, by Theorem 2, no more than one point is common to any two lines.

Theorem 4. There is at least one line containing any two points.
Proof: This statement is establihsed directly by Axiom 4.

Theorem 5. Not all lines in the system pass through the same point.

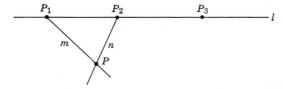

Figure 9.14

Proof: Assume that all lines pass through a common point. By Axiom 1, there exists a line l and, by Axiom 2, this line contains exactly three points, P_1, P_2, and P_3. One of these three points must be the common point, since line l must pass through it. By Axiom 3, there is a point P not on l (Figure 9.14) and, by Axiom 4, the points P and P_1 determine a unique line m. Then neither P_2 nor P_3 can lie on line m, by Theorem 3, and, therefore, neither can be the common point for all lines. By a similar argument, line n is determined by P and P_2, and neither P_1 nor P_3 can be a common point for all lines. Therefore, the assumption that there is a common point for all lines in untenable, and the theorem is established.

Theorem 6. Three and only three lines pass through each point.

Proof: Let point P be any point in the system. Then, by Theorem 4, there is a line l which does not contain P. Let P_1, P_2, P_3 be the three points on l. By Axiom 4, these points determine the lines m, n, and k illustrated in Figure 9.15. Therefore, at least three lines pass through the point P. To prove that no more than three lines pass through P, consider a fourth line q through P. Then q must have a point in common with l. By Axiom 4, this point is distinct from P_1, P_2, or P_3. But this contradicts Axiom 2. Therefore, only three lines pass through the point P.

These six theorems establish the duality of the five given axioms. One important consequence of this property is that every subsequent theorem can be dualized to produce a valid theorem. Another interesting consequence is that the system of finite geometry could have been structured with equal validity by starting with a set of undefined elements called lines and subsets of lines called points. This serves to emphasize the abstract nature of finite geometry.

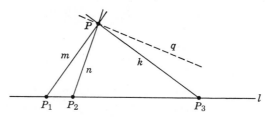

Figure 9.15

EXERCISE 49

FINITE GEOMETRY

Given seven members of a club—Bob, Jim, Paul, Dick, John, Fred, and Tom.

1. Determine two different sets of committees such that each committee will consist of three members, any two committees will have one member in common, and any two members will serve together on a single committee.
2. Determine a set of three-man committees such that each man serves on exactly one committee with every other man.
3. Show that the committees formed in Problems 1 and 2 are equivalent.

The following statements will develop a proof that the five axioms listed for a finite geometry will result in a system of exactly seven points and seven lines. Prove each statement by citing an appropriate axiom or proved theorem. The notations refer to Figure 9.16.

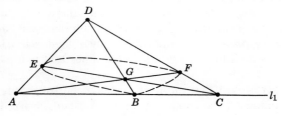

Figure 9.16

4. There exists a line l_1.
5. The line l_1 contains exactly three points, namely, A, B, and C.
6. There is a point D not on l_1.
7. D and A determine a line l_2.
8. l_2 must contain a third point, E.
9. E is distinct from A and D.
10. E is distinct from B and C, that is, E is a fifth point in the system.
11. D and B determine a line l_3.
12. l_3 must contain a third point, G.
13. G is distinct from B and D.
14. G is distinct from A, C, and E, that is, G is a sixth point in the system.
15. D and C determine a line l_4.
16. l_4 must contain a third point, F.
17. F is distinct from D and C.
18. F is distinct from A, B, E, and G, that is, F is a seventh point in the system.

19. The subset of points A, G, F determine a line l_5.
20. The subset of points E, G, C determine a line l_6.
21. The subset of points E, B, F determine a line l_7.
22. If there is an eighth point, H, it cannot lie on l_1.
23. Then H determines a line with each of the points A, B, C.
24. None of these lines can contain D.
25. H and D must determine a line.
26. But four lines cannot pass through D.
27. Therefore, there cannot exist an eighth point H.

50. TRANSFINITE NUMBERS

Each of us has, at some time, wondered about the nature of infinity. The concept of something being endless is difficult to conceive, yet very early in our mathematical experience we recognized that there is no largest natural number. Later, we encountered the notion of infinity again in considering such physical concepts as space and time or such geometric concepts as the number of points on a line or the number of lines that can pass through a point.

Experience has proved that intuition is not to be trusted in dealing with the concept of infinity. Even competent mathematicians, up to the year 1872, erroneously conceived infinity to be a single magnitude which could be applied indiscriminately to either the number of elements in the set of natural numbers or the number of points on a line. The first definitive work dealing with the mathematics of infinitely large numbers was developed by Georg Cantor in a series of bold, imaginative articles beginning in 1872. Our purpose is to introduce briefly and informally some of the concepts that have resulted from Cantor's work.

Cardinality and Equivalent Sets

Counting, as we have indicated earlier in this text, is basically a process of setting up a one-to-one correspondence between the things we wish to count and the ordered elements of the set of natural numbers. For example, to count the points of direction on a compass we set up the correspondence:

N	NbE	NNE	NEbN	NE	NEbE	ENE	EbN	E	⋯	NNW	NbW
↕	↕	↕	↕	↕	↕	↕	↕	↕		↕	↕
1	2	3	4	5	6	7	8	9	⋯	31	32

Thus, the number of points is 32, which we call the *cardinality* of the set of points. We also refer to 32 as the *cardinal number* of the set. In general, a set S is said to have a cardinal number n if and only if the elements of S can be put in one-to-one correspondence with the set of natural numbers $\{1, 2, 3, \ldots n\}$.

If two sets have the same cardinal number, they are said to be *equivalent*. Thus we link together three basic notations:

> one-to-one correspondence
> cardinality
> equivalence

Our preoccupation with these seemingly simple concepts is, of course, to develop a basis for testing the equivalence of infinite sets.

Infinite Sets

A meaningful discussion of infinite sets should probably begin with some attempt to draw a clear distinction between finite and infinite sets. One approach is to define these terms as follows.

Definition 1. A finite set is one which can be placed in one-to-one correspondence with a subset of the set $\{1, 2, 3, \ldots n\}$ where n is a natural number. An infinite set is one which is not finite.

The subsets of a finite set $\{1, 2, 3, \ldots n\}$, as illustrated in Section 7, number 2^n. For example, there are 2^{10} or 1024 subsets which can be formed from the elements of the set $\{1, 2, 3, \ldots 10\}$ and, according to the definition, each of these will be a finite set. The matter of infinite sets is dealt with indirectly. Although the definition of infinity is mathematically acceptable, it is difficult to apply. For example, to prove that the set $\{3, 5, 7, 9, \ldots\}$ is infinite would require a demonstration that no equivalent set has finite cardinality. An alternative definition approaches the matter of infinity directly.

Definition 2. An infinite set is one which can be put in one-to-one correspondence with a proper subset of itself. A finite set is one which is not infinite.

EXAMPLE 1. Prove that the set of natural numbers $N = 1, 2, 3, \ldots$ is infinite.

Solution: Choose a nonterminating proper subset, say, $A = 4, 5, 6, \ldots$. To prove that N is infinite we have to establish a one-to-one correspondence between the elements of N and its proper subset A. We do this with the following display:

$$\begin{array}{cccccccc} 1 & 2 & 3 & 4 & 5 & \cdots & n & \cdots \\ \updownarrow & \updownarrow & \updownarrow & \updownarrow & \updownarrow & & \updownarrow & \\ 4 & 5 & 6 & 7 & 8 & \cdots & n+3 & \cdots \end{array}$$

Since for any element of set N there exists one and only one corresponding element in set A, the two sets are equivalent. Therefore N is infinite.

The First Transfinite Cardinal

The cardinal number assigned to the set of natural numbers is $\aleph_0$, which is read "aleph-null." It is the first of the so-called transfinite (beyond the finite) numbers. Of course, many other proper subsets of N can be placed in one-to-one correspondence with the set N and, therefore, are equivalent sets. Among the most noteworthy of these are the even numbers, odd numbers, squares of natural numbers, prime numbers, and the natural numbers from any finite number on. Thus, the cardinality of each of these infinite sets is $\aleph_0$, which leads to some interesting and unexpected properties of the number $\aleph_0$.

For example, if we consider the two disjoint sets

$K = \{\text{first } n \text{ natural numbers}\}$, where n is a finite number

$M = \{\text{remaining natural numbers beginning with } n + 1\}$

then the union of K and M is the set of natural numbers. But the cardinality of set M, by setting up a correspondence as in Example 1, can be shown to be $\aleph_0$. We know that the cardinality of set K is n and that of set N is $\aleph_0$. Therefore, for any natural number n,

$$\aleph_0 + n = \aleph_0 \tag{1}$$

Similarly, the set of even numbers is of size $\aleph_0$, as is also the set of odd numbers. But these are disjoint sets, and their union is the set of natural numbers, whose size is also $\aleph_0$. Therefore we are led to conclude that

$$\aleph_0 + \aleph_0 = \aleph_0 \tag{2}$$

If transfinite numbers are to obey the associative law, then

$$(\aleph_0 + \aleph_0) + \aleph_0 = \aleph_0 \tag{3}$$

and, in general,

$$\aleph_0 + \aleph_0 + \aleph_0 + \cdots \aleph_0 = \aleph_0 \tag{4}$$

It can be further shown, by setting up more elaborate systems of correspondence, that

$$n \cdot \aleph_0 = \aleph_0 \tag{5}$$

and

$$\aleph_0 \cdot \aleph_0 = \aleph_0 \tag{6}$$

The ubiquitous nature of the number $\aleph_0$ in all these considerations justifies a natural curiosity as to whether we can ever determine a larger transfinite number. To test this we now examine some infinite sets which appear to extend beyond the set of natural numbers.

Other Transfinite Cardinals

We examine first the cardinality of the set of integers—a set which certainly appears to be more extensive than the set of natural numbers, since it includes the negative integers and zero in addition to the natural numbers. The question is easily settled by means of the following correspondence between the elements of the set N of natural numbers and the set I of integers:

Elements of N: 1   2   3   4   5   6   7 $\cdots$
$\updownarrow \;\; \updownarrow \;\; \updownarrow \;\; \updownarrow \;\; \updownarrow \;\; \updownarrow \;\; \updownarrow$
Elements of I: 0   1 -1 2 -2 3 -3 $\cdots$

For every element n of the set N, there is one and only one corresponding element i of the set I, related through the function

$$i = \begin{cases} \frac{1}{2}n & \text{if } n \text{ is even} \\ \frac{1}{2}(1 - n) & \text{if } n \text{ is odd} \end{cases}$$

Therefore, the set of integers is equivalent to the set of natural numbers, and its cardinality is $\aleph_0$.

The investigation of the cardinality of the next enlargement of the set N—the set of rationals—requires somewhat more ingenuity. Ingenuity is required because there is an infinite number of rational numbers between any two rational numbers, and some systematic way of displaying them must be established if we are to test the existence of a correspondence with the set of natural numbers. Cantor devised an arrangement, which is shown in Figure 9.17, to prove that the set of rationals is equivalent to the set of natural numbers. The arrows in Figure 9.17 show the order of counting the rational numbers—the fact that each rational may appear in the arrangement an unlimited number of times is indicated by the circled numbers. These circled numbers may be disregarded, although this does not affect the final conclusion. The one-to-one correspondence can now be listed as follows:

1   2   3   4   5   6   7   8   9   10   11 $\cdots$
$\updownarrow \; \updownarrow \; \updownarrow \; \updownarrow \; \updownarrow \; \updownarrow \; \updownarrow \; \updownarrow \; \updownarrow \; \updownarrow \; \updownarrow$
$\frac{1}{1} \; \frac{2}{1} \; \frac{1}{2} \; \frac{1}{3} \; \frac{3}{1} \; \frac{4}{1} \; \frac{3}{2} \; \frac{2}{3} \; \frac{1}{4} \; \frac{1}{5} \; \frac{5}{1} \; \cdots$

Since every positive rational number must ultimately appear in the schematic diagram of Figure 9.17 once and can be associated with a natural number in the sequence indicated by the diagonals once and only once, we have established an equivalence of the set of positive rationals and natural numbers. We need only to adjoin the negative rationals and zero to complete the task. This is easily done.

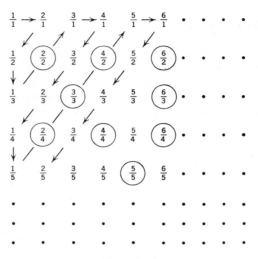

Figure 9.17

If we designate the positive rationals in the order generated in Figure 9.17 as $r_1, r_2, r_3, \ldots$, then we can form the set R consisting of *all* rationals as follows:

$$R = \{0, r_1, -r_1, r_2, -r_2, r_3, -r_3, \ldots\}$$

This set can clearly be placed in one-to-one correspondence with the natural numbers, so that the set R also has cardinality $\aleph_0$.

It begins to appear as though each enlargement of the set N results in a set whose cardinality is still $\aleph_0$. This, however, would be a dangerous induction—actually there is an infinite number of transfinite numbers greater than $\aleph_0$. We now demonstrate the existence of one of these transfinites—a number associated with the subset of real numbers between any two finite numbers. For specificity we will consider the real numbers in the interval between 0 and 1.

Every real number is either rational or irrational. As discussed in Section 15, each irrational number is expressible as a nonterminating decimal, and certain rationals (such as $7/33 = 0.212121\cdots$) can be expressed as nonterminating decimals. We now agree to write each remaining rational as a nonterminating decimal, as $1/8 = 0.125000\cdots$ or $1/8 = 0.124999\cdots$. With these agreements we now conclude that each real number between 0 and 1 can be written as a nonterminating decimal of the form

$$0.a_1 a_2 a_3 a_4 a_5 \cdots$$

with each a_i designating one of the digits 0, 1, 2, 3, 4, 5, 6, 7, 8, 9.

50. Transfinite Numbers

Let us now assume that each real number in the given interval can be arranged to correspond to a single natural number according to the arrangement in Figure 9.18.

$$
\begin{aligned}
1 &\leftrightarrow 0.a_1 a_2 a_3 a_4 \cdots \\
2 &\leftrightarrow 0.b_1 b_2 b_3 b_4 \cdots \\
3 &\leftrightarrow 0.c_1 c_2 c_3 c_4 \cdots \\
4 &\leftrightarrow 0.d_1 d_2 d_3 d_4 \cdots \\
\cdot &\leftrightarrow \cdot \quad \cdot \quad \cdot \quad \cdot \quad \cdot \quad \cdot \quad \cdot \quad \cdot \\
\cdot &\leftrightarrow \cdot \quad \cdot \quad \cdot \quad \cdot \quad \cdot \quad \cdot \quad \cdot \quad \cdot \\
\cdot &\leftrightarrow \cdot \quad \cdot \quad \cdot \quad \cdot \quad \cdot \quad \cdot \quad \cdot \quad \cdot
\end{aligned}
$$

Figure 9.18

If this were possible, then the subset of real numbers would be equivalent to the set of natural numbers. But it is not possible because we can always find real numbers which differ from every number in the listing and, therefore, cannot be placed in correspondence with a natural number. To illustrate this, consider the specific array of real numbers

$$0.132458 \cdots$$

$$0.223726 \cdots$$

$$0.185940 \cdots$$

$$0.332781 \cdots$$

To determine a real number different from any in the listing we merely choose a first digit different from 1, a second digit different from 2, a third digit different from 5, a fourth digit different from 7, and so on. The resulting decimal number will differ in at least one place from any in the set.

Thus, the cardinality of the subset of real numbers is greater than $\aleph_0$. It is usually designated by the transfinite number **c**, called the *cardinality of the continuum*. Since the points on a line segment of unit length can be placed in one-to-one correspondence with the real numbers in the interval 0 to 1, it follows that the cardinality of the set of points on such a line segment is **c**, and it can be further demonstrated (see Section 7) that the cardinality of the set of points on a line segment of *any* length is **c**.

Cantor's work raised as many questions as it settled. For example, it has been proved that $\aleph_0$ is the smallest transfinite number and that there exists an infinite number of transfinite numbers. But despite great efforts, no one has been able to prove the widely held conjecture that **c** follows $\aleph_0$ directly in size order of these numbers, that is, that there is no transfinite number greater than $\aleph_0$ but less than c. (In recent years it has been shown that this conjecture is independent of the axioms of set theory and, therefore, cannot be deduced from them.) Further, the procedure known as Cantor's diagonal process, illustrated in Figure 9.18, has not been completely accepted by all mathematicians, and the entire treatment of infinity on the basis of set theory has resulted in a number of paradoxes.

It is perhaps appropriate to close our inquiry into mathematics on this note—an indication that modern mathematics is not final and complete but that it is undergoing constant reexamination as it reaches ever higher levels of abstraction and develops ever more sophisticated ways of solving problems.

EXERCISE 50

Transfinite Numbers

1. Personal income in the United States in a recent year was at the $670 billion level. Approximate the number of years it would take to count to this number at the rate of one number per second.
2. Show that there are as many multiples of seven as there are natural numbers.
3. Show that the cubes of natural numbers (1, 8, 27, 64, ...) are equivalent to the set of natural numbers.

The correspondence between the integers and the natural numbers was expressed by the function

$$i = \begin{cases} \frac{1}{2} n \text{ if } n \text{ is even} \\ \frac{1}{2} (1 - n) \text{ if } n \text{ is odd} \end{cases}$$

Find the element which corresponds to:

4. $n = 126$
5. $n = 101$
6. $n = 355$
7. $i = 18$
8. $i = -37$
9. $i = 29$

Expand the arrangement illustrated in Figure 9.17 to determine the rational number that corresponds to each of the following natural numbers (Problems 10–12).

10. 17
11. 25
12. 30

13. Using a diagram, show that there are as many points on a semicircle as there are on a line of infinite length.
14. Prove that the set of all odd numbers has cardinality $\aleph_0$.
15. Prove that the cardinality of all rationals with denominator 5 is $\aleph_0$.

16. Prove that $\aleph_0 + 8 = \aleph_0$.
17. Prove that the sets $\{1, 4, 9, 16, 25, 36, \ldots\}$ and $\{1, 8, 27, 64, 125, 216, \ldots\}$ are equivalent.
18. (a) Can you always find at least one rational number between any two given rational numbers? (b) Show that this implies the existence of an infinite set of rationals between any two given rationals. (c) What is the cardinality of this set?

Appendix

Table of Squares and Square Roots

No.	Square	Square root	No.	Square	Square root	No.	Square	Square root
1	1	1.000	34	1,156	5.831	67	4,489	8.185
2	4	1.414	35	1,225	5.916	68	4,624	8.246
3	9	1.732	36	1,296	6.000	69	4,761	8.307
4	16	2.000	37	1,369	6.083	70	4,900	3.367
5	25	2.236	38	1,444	6.164	71	5,041	8.426
6	36	2.449	39	1,521	6.245	72	5,184	8.485
7	49	2.646	40	1,600	6.325	73	5,329	8.544
8	64	2.828	41	1,681	6.403	74	5,476	8.602
9	81	3.000	42	1,764	6.481	75	5,625	8.660
10	100	3.162	43	1,849	6.557	76	5,776	8.718
11	121	3.317	44	1,936	6.633	77	5,929	8.775
12	144	3.464	45	2,025	6.708	78	6,084	8.832
13	169	3.606	46	2,116	6.782	79	6,241	8.888
14	196	3.742	47	2,209	6.856	80	6,400	8.944
15	225	3.873	48	2,304	6.928	81	6,561	9.000
16	256	4.000	49	2.401	7.000	82	6,724	9.055
17	289	4.123	50	2,500	7.071	83	6,889	9.110
18	324	4.243	51	2,601	7.141	84	7,056	9.165
19	361	4.359	52	2,704	7.211	85	7,225	9.220
20	400	4.472	53	2,809	7.280	86	7,396	9.274
21	441	4.583	54	2,916	7.348	87	7,569	9.327
22	484	4.690	55	3,025	7.416	88	7,744	9.381
23	529	4.796	56	3,136	7.483	89	7,921	9.434
24	576	4.899	57	3,249	7.550	90	8,100	9.487
25	625	5.000	58	3,364	7.616	91	8,281	9.539
26	676	5.099	59	3,481	7.681	92	8,464	9.592
27	729	5.196	60	3,600	7.746	93	8,649	9.644
28	784	5.292	61	3,721	7.810	94	8,836	9.695
29	841	5.385	62	3,844	7.874	95	9,025	9.747
30	900	5.477	63	3,969	7.937	96	9,216	9.798
31	961	5.568	64	4,096	8.000	97	9,409	9.849
32	1,024	5.657	65	4,225	8.062	98	9,604	9.899
33	1,089	5.745	66	4,356	8.124	99	9,801	9.950

Answers to Odd-Numbered Problems

EXERCISE 1

1. Do not agree.
3. Agree.
5. Agree.
7. Agree.
9. Agree.
11. Not true.
13. Not true. $11 + 1.1 = 11 \times 1.1$.
15. Not true.
17. Yes.
19. $S + P$ and $S - P$ are perfect squares.
21. If 1 is subtracted from the square of an odd number greater than 1, the result is divisible by 8.
23. (a) $6^3 = 31 + 33 + 35 + 37 + 39 + 41.$
    $7^3 = 43 + 45 + 47 + 49 + 51 + 53 + 55.$
    $8^3 = 57 + 59 + 61 + 63 + 65 + 67 + 69 + 71.$
    $9^3 = 73 + 75 + 77 + 79 + 81 + 83 + 85 + 87 + 89.$
    (b) The cube of every odd number, n, may be expressed as the sum of n consecutive odd numbers, beginning with $n^2 - n + 1$.

EXERCISE 2

1. Valid and true.
3. Not valid, may or may not be true.
5. Not valid, may or may not be true.

EXERCISE 2—continued

7. Not valid, true.
9. Not valid and not true.
11. Valid and true.
13. Not valid, true.
15. Valid, may or may not be true.
17. Not valid, may or may not be true.
19. Smith has a great deal of imagination.
21. No conclusion.
23. No conclusion.
25. No conclusion.
27. No conclusion.
29. Mary is over 21.
31. Some problems are worthwhile.
33. Some students eventually graduate.

EXERCISE 3

1. An even number is a number which is divisible by 2.
3. A triangle is a polygon with three sides.
5. A proportion is an equality of two ratios.
7. A binomial is a polynomial consisting of two terms.
9. A median of a triangle is a straight-line segment drawn from any vertex to the midpoint of the opposite side.
11. A prime number is a number which has no factors except 1 and the number itself.
13. An identity is an equation which is true for all values of the variable.
15. A diameter of a circle is a chord which passes through the center of the circle.
17. The definition fails to give the meaning of an exponent.
19. The definition is too general; it does not indicate the particular kind of an equation.
21. The word to be defined is used in the definition.
23. The definition is not accurate. Negative whole numbers are less than 1.
25. The word "warm" is used in two different senses.
27. The thermometer may have indicated the temperature of a freezing solution in a warm room.
29. The word "cabinet" is used with two different meanings.
31. The product $k(k + 1)(k + 2)(k + 3)$, plus 1, equals $k^4 + 6k^3 + 11k^2 + 6k + 1$. This expression may be written as $(k^2 + 3k + 1)^2$, that is, a perfect square.
33. $n^2 - n + 1 = n(n - 1) + 1$.
   If n is odd, $n - 1$ is even, and if n is even, $n - 1$ is odd.

EXERCISE 3—continued

Since $n(n - 1)$ is therefore the product of one odd number and one even number, the product is always an even number.
Therefore $n(n - 1) + 1$ is always an odd number.
Therefore $n^2 - n + 1$ is always an odd number when n is an integer.

EXERCISE 4

1. Statement, T
3. Statement, T or F.
5. Statement, T or F.
7. Statement, T.
9. Not a statement.
11. Open sentence, 6.
13. Open sentence, T if $x = 5$.
15. Statement, T or F.
17. Open sentence.
19. $A \wedge B$.
21. $\sim A \wedge B$.
23. $\sim A \wedge \sim B$.
25. I am going to college and I will get a job.
27. I am not going to college and I will get a job.
29. I am not going to college and I will not get a job.
31. It is false that I will both go to college and get a job.
33. The exclusive disjunction.
35. T.
37. F.
39. F.
41. F.
43. T.
45. F.
47. F.

EXERCISE 5

1. $B \to C$.
3. $\sim A \to \sim C$.
5. $A \leftrightarrow C$.
7. $(A \wedge B) \to C$.
9. $(B \wedge A) \leftrightarrow C$.
11. $\sim C \leftrightarrow \sim B$.
13. T.
15. F.
17. T.
19. T.
21. T.
23. T.
25. If the sum of two numbers is even, the numbers are even.
    If two numbers are not even, their sum is not even.
    If the sum of two numbers is not even, the numbers are not even.
27. If the product of two numbers is odd, the numbers are odd.
    If two numbers are not odd, their product is not odd.
    If the product of two numbers is not odd, the numbers are not odd.
29. If a number has two equal factors, it is a perfect square.
    If a number is not a perfect square, it does not have two equal factors.
    If a number does not have two equal factors, it is not a perfect square.

EXERCISE 5—continued

31. A necessary and sufficient condition that one root of the quadratic equation $ax^2 + bx + c = 0$ is zero is that $c = 0$.

33. $q \rightarrow p$. **35.** $p \rightarrow q$. **37.** $\sim p \rightarrow \sim q$.

39. $q \leftrightarrow p$.

EXERCISE 6

1. Valid. **3.** Not valid. **5.** Not valid.

7. Valid. **9.** Valid. **11.** Valid.

13. Valid. **15.** Not valid. **17.** No.

19. Yes. **21.**

p	$\sim p$	$p \vee \sim p$
T	F	T
F	T	T

23. Valid.

25. (a) $\{[(p \rightarrow q) \wedge (q \rightarrow r)] \wedge \sim r\} \rightarrow \sim p$.
(b) *Hint:* Does $p \rightarrow r$? Is r a necessary or sufficient condition for p? If r is false, what can be said about p?

EXERCISE 7

1. (a) $\{1, 2, 3, 4, 5, 6, 7, 8, 9, 10\}$. (b) $\{2, 4, 6, 8, 10\}$. (c) $\{1, 3, 5, 7, 9\}$.
(d) $\{2, 5\}$.

3. (a) $\{1, 2, 3, 4, 5\}$. (b) $\{0, 1, 2, 3\}$. (c) $\{3, 4\}$. (d) $\{0, 1, 2, 3, 4\}$. (e) $\{4\}$.
(f) $\{\ \}$.

5. (a) $A = \{x \mid x$ is an even number, and $2 \leq x \leq 10\}$.
(b) $B = \{x \mid x$ is an odd number, and $1 \leq x \leq 9\}$.
(c) $C = \{x \mid x$ is a perfect square, and $1 \leq x \leq 25\}$.
(d) $D = \{x \mid x$ is a perfect cube, and $1 \leq x \leq 125\}$.
(e) $E = \{x \mid x$ is a number ending in 4, and $4 \leq x \leq 44\}$.
(f) $F = \{x \mid x$ is the reciprocal of a natural number from 2 to 9 inclusive$\}$.

7. $-1, -2, -3, -4, -5$.

9. $\bar{A} = \{b, d, e, g\}$. $\bar{B} = \{c, d, e\}$.

11. $\{\ \}$ **13.** F. **15.** T. **17.** $\{\ \}$.

19. Yes. $3 + (-4) = -1$, an element of B.

21. Yes. $(2)(-2) = -4$, an element of B.

23. No. For all values of y, y^2 is 0 or a positive number.

Answers to Odd-Numbered Problems

EXERCISE 8

1. $\{0, 1, 2, 3, 4\}$.
3. $\{1, 2, 3, 4, 5\}$.
5. $\{0, 1, 2, 3, 5\}$.
7. $\{0, 4, 5\}$.
9. $\{5\}$.
11. $\{4\}$.
13. $\{0\}$.
15. $\{1, 2, 3\}$.
17. $\{0, 1, 2, 3\}$.
19. $\{1, 2, 3, 4, 5, 6\}$.
21. $\{1, 2\}$.
23. $7, 8, \ldots$.
25. U.
27. A.
29. L_2.
31. A single point.
33. I is a point not in U.
35. The point of tangency.
37. The set A.

EXERCISE 9

1. No.
3. Yes.
5. Yes.
7. No.
9. Yes.
11. CCLVII.
13. MCMLXVI.
15. XX.
17. CLXV.
19. CXCVI.
21. The last date is 51.
25. $4 \times 6^2 + 3 \times 6 + 2 = 164_{10}$.
27. $3 \times 6^2 + 5 \times 6 + 4 = 142_{10}$.
29. $4 \times 6^2 + 1 \times 6 + 2 = 152_{10}$.
31. 215_{10}.
33. (a) 1295_{10}. (b) 1296_{10}. (c) 9331_{10}.
35. 203.
37. 1033.
39. 505.
41. 43.
43. 1101.
45. 143.
47. 5442.
49. 1422.
51. 3254.
53. 420.
55. 103432.

EXERCISE 10

1. 5.
3. 27.
5. 651.
7. 7589.
9. $(100100)_2$.
11. $(11101)_2$.
13. $(1011011)_2$.
15. $(1111111)_2$.
17. $(1001\ 0110)_{bcd}$.
19. $(0010\ 0010\ 1000)_{bcd}$.
21. $(0111\ 0001\ 0100\ 0010)_{bcd}$.
23. $(0010\ 0111\ 0110\ 0011)_{bcd}$.
25. $(11000)_2$.
27. $(100111)_2$.
29. $(110111)_2$.
31. $6 \times 8 + 3 \times 5 = 51_{\text{ten}}$.
33. $2 \times 8^2 + 0 \times 8 + 6 = 134_{\text{ten}}$.
35. $5 \times 8^2 + 7 \times 8 + 0 = 376_{\text{ten}}$.
37. $6 \times 8^2 + 2 \times 8 + 1 = 401_{\text{ten}}$.
41. $(001\ 100\ 011\ 010)_{bco}$.
43. $(111\ 110\ 011)_{bco}$.

EXERCISE 11

1. $1 = 1$. **3.** $7 = 7$. **5.** $4 = 9 + 7 = 4$.
7. $8 = 8$. **9.** $4 = 4$. **11.** Not commutative.
13. Not commutative. **15.** Not associative. **17.** Associative.
19. $2 = 2$. **21.** $1 = 1$. **23.** $5 = 5$. **25.** $3 = 3$.
27. $2 = 2$. **29.** $4 = 4$.

31.

+	1	2	3	4	5
1	2	3	4	5	1
2	3	4	5	1	2
3	4	5	1	2	3
4	5	1	2	3	4
5	1	2	3	4	5

33.

×	1	2	3	4	5
1	1	2	3	4	5
2	2	4	1	3	5
3	3	1	4	2	5
4	4	3	2	1	5
5	5	5	5	5	5

35. $2 + 4 = 1, 4 + 4 = 3$, or equivalent answers.

37. $4(3 + 4) = 4 \times 3 + 4 \times 4 \quad 2(2 + 3) = 2 \times 2 + 2 \times 3$
$\qquad 4 \times 2 = 2 + 1 \qquad\qquad 2 \times 5 = 4 + 1$ or equivalent
$\qquad\quad 3 = 3 \qquad\qquad\qquad\quad 5 = 5$ answers.

39. $3 \times 1 = 3$
$\quad\; 4 \times 1 = 4$ or equivalent answers.

41. None. **43.** $\square = x$, $\bigcirc = +$.

45. (a)

*	i	h	v	r
i	i	h	v	r
h	h	i	r	v
v	v	r	i	h
r	r	v	h	i

(b) Yes, since each operation gives an element of the set.
(c) $i * v = v \quad v * r = h$ The operation $*$ is
$\quad v * i = v \quad r * v = h$ commutative.
(d) $(h * i) * v = h * v = r$
$\quad h * (i * v) = h * v = r$
$\quad (v * h) * r = r * r = i$ The operation $*$ is
$\quad v * (h * r) = v * v = i$ associative.
(e) Yes. The element i.

EXERCISE 12

1. 3. **3.** 13. **5.** 11. **7.** Is.
9. Is not. **11.** Is not. **13.** Is not. **15.** Is.
17. $1 + 3 = 1 + 2' = (1 + 2)' = 3' = 4$.
19. $2 \times 3 = 2 \times 2' = 2 \times 2 + 2 = 2 \times 1' + 2 = (2 \times 1 + 2) + 2$
$\qquad\quad = (2 + 2) + 2 = 4 + 2 = 6$.

EXERCISE 12—continued

21. Using procedure in Example 2, show that $3 \times 2 = 6$ and $2 \times 2 = 4$. Using procedure in Example 1, show that $6 + 4 = 10$.
23. Multiplication by 1.　　**25.** Addition and multiplication.
27. Closed.　　**29.** Not closed.
31. Not closed.　　**33.** 715.
35. 3176.　　**37.** 27.
39. 8.　　**41.** 135.
43. (a) Expanded notation. (b) Distributive law. (c) Commutative law of multiplication. (d) Distributive law. (e) Associative law of addition. (f) Expanded notation. (g) Associative law of addition. (h) Decimal notation.

EXERCISE 13

1. 0.　　**3.** 0.　　**5.** No meaning.　　**7.** No meaning.　　**9.** 0.
11. $(0 + a) + b = a + b$.　　**13.** $0 \times (a + b) = 0$.
　　$0 + (a + b) = a + b$.　　　　$0 \times a + 0 \times b = 0 + 0 = 0$.
15. 2.　　**17.** 4.　　**19.** -4.　　**21.** 0.　　**23.** -17.　　**25.** 0.　　**27.** Is not.
29. Is.　　**31.** Is not.　　**33.** $(9, 15)$.　　**35.** $(40, 40)$.　　**37.** $(8, 8)$.
39. $(12, 13)$.　　**41.** 2.　　**43.** -5.　　**45.** -5.　　**47.** -11.　　**49.** -32.
51. 20, 75, 10, 3.　　**53.** 8, -48, 16, -3.　　**55.** -12, -108, -24, -3.
57. -24, 128, -8, 2.　　**59.** 0, -9, -6, -1.　　**61.** -8, 16, 0, 1.
63. 6, 0, -6, 0.　　**65.** -12, 32, -4, 2.　　**67.** $(0, b) + (b, 0) = (b, b) = 0$, $(b, 0)$ is therefore the additive inverse of $(0, b)$.
69. $(0, b) + (0, c) = (0, b + c)$. $(0, b + c)$ is a negative integer.
71. $(0, b) \times (0, c) = (0 + bc, 0 + 0) = (bc, 0)$. $(bc, 0)$ is a positive integer.

EXERCISE 14

1. Not equal.　　**3.** Equal.　　**5.** Equal.　　**7.** $(13, 6)$.
9. $(6, 35)$.　　**11.** $(-13, 45)$.　　**13.** $(19, 42)$.　　**15.** $(9, 32)$.
17. $(7, 3) \times (2, 5) = (14, 15)$.　　**19.** $(-4, 13) + (-1, 2) = (-21, 26)$.
　　$(2, 5) \times (7, 3) = (14, 15)$.　　　　$(-1, 2) + (-4, 13) = (-21, 26)$.
21. $(3, 2) \times [(1, 2) + (5, 7)] = (51, 28)$.
　　$(3, 2) \times (1, 2) + (3, 2) \times (5, 7) = (51, 28)$.

EXERCISE 14—continued

23. Eight of the three equal parts of a quantity. Eight divided by three. A ratio of eight to three.

25. Two of nine equal parts of a quantity. Two divided by nine. A ratio of two to nine.

27. 4/14, 6/21, 8/28, 10/35, or equivalent answers.

29. 18/16, 27/24, 36/32, 45/40, or equivalent answers.

31. 11/4.   **33.** 17/24.   **35.** 1/5.   **37.** 45/14.   **39.** (a) $(-4, 5)$. (b) $(5, 4)$.

41. (a) $(-8, 1)$. (b) $(1, 8)$.   **43.** (a) $-3/4$. (b) $4/3$.

45. (a) 8/3. (b) $-3/8$.   **47.** $a/c + b/c = \dfrac{ac + bc}{cc} = \dfrac{c(a + b)}{cc} = \dfrac{a + b}{c}$.

49. 7/8.   **51.** 3/4.   **53.** 11/15.   **55.** 34/35.   **57.** 47/30.   **59.** 319/210.

EXERCISE 15

1. Z, R, I, W, N.   **3.** Z, R.   **5.** $Z, \bar{R}$.   **7.** Z, R.   **9.** Z, R, I, W, N.

11. $0.4166\ldots$ **13.** $0.7777\ldots$ **15.** $1.6666\ldots$ **17.** 13/100.

19. 43/9.   **21.** 5866/999.   **23.** Will neither terminate nor repeat.

25. Will not terminate but will repeat.   **27.** Will not terminate but will repeat.

29. 1.23.   **31.** 3.65.   **33.** 36.5.   **39.** (a) 2.318 and 2.319, or equivalent answers. (b) An infinite number. (c) $2.317171171117111\ldots$, $2.315253545556\ldots$. (d) An infinite number.

EXERCISE 16

1. None.   **3.** None.   **5.** All.   **7.** Transitive.   **9.** All.

11. Problems 5, 6, 9.   **13.** Axiom of symmetry.   **15.** Axiom of transitivity.

17. Axiom of addition.   **19.** Axiom of multiplication.   **21.** Axiom of addition.

23. If John is a brother of Joe, Joe is a brother of John.   **25.** Mary likes herself.

27. If John is the father of Joe, and Joe is the father of Mary, John is not the father of Mary.

EXERCISE 17

1. Do not exist.   **3.** $-2/3$ and $3/2$.   **5.** True. $2 + 1 = 1$, $1 + 1 = 2$, or equivalent answers.   **7.** True. $2 + 1 = 0$, $1 + 2 = 0$, or equivalent answers.

Answers to Odd-Numbered Problems   313

EXERCISE 17—continued

9. True. $2 + (1 + 1) = 2 + 2 = 1$, $(2 + 1) + 1 = 0 + 1 = 1$, or equivalent answers.

11. True. $2(1 + 1) = 2 \times 2 = 1$, $2(1 + 1) = 2 \times 1 + 2 \times 1 = 2 + 2 = 1$, or equivalent answers.

13. True. The identity element for multiplication is 1. $1 \times 1 = 1$, $2 \times 2 = 1$, etc.

15. True. $1 \times 1 = 1, 2 \times 2 = 1$.

17.

+	0	1	2	3	4	5
0	0	1	2	3	4	5
1	1	2	3	4	5	0
2	2	3	4	5	0	1
3	3	4	5	0	1	2
4	4	5	0	1	2	3
5	5	0	1	2	3	4

×	0	1	2	3	4	5
0	0	0	0	0	0	0
1	0	1	2	3	4	5
2	0	2	4	0	2	4
3	0	3	0	3	0	3
4	0	4	2	0	4	2
5	0	5	4	3	2	1

This system is not a field. At least one axiom does not hold—there is no multiplicative inverse.

19. No. There is no additive inverse for $x + \sqrt{2}$.

EXERCISE 18

1. $2a^2b^3$.   3. $16x^3y^2$.   5. $-4pq^2r$.
7. $x^3 \cdot x^5 = x \cdot x \cdot x \cdot x \cdot x \cdot x \cdot x \cdot x = x^8$.   9. $(r^2)^3 = r^2 \cdot r^2 \cdot r^2 = r^6$.
11. $(b^m)^3 = b^m \cdot b^m \cdot b^m = b^{3m}$.   13. $7a + 3c$.   15. $5u^3 - 5u^2v + 4uv$.
17. $9x^2 + 8y^2$.   19. $2a^2 - a - 3$.   21. $-11q^2 - 5pq$.   23. $2a^2 + 6a + 5$.
25. $-5a - 2b$.   27. $y^2 - 10y + 21$.   29. $3v^2 + 16v - 12$.   31. $49b^2 - 1$.
33. $9x^2 - 12x + 4$.   35. $24a^2 - 47ab - 21b^2$.   37. No.   39. $3a + 2$.
41. $8x^2 + 2h^2$.   43. $2hx + h^2 + 2h$.   45. $4x + 2h + 5$.   47. $a^2(a - 1)$.
49. $\pi r(r + 2)$.   51. $2(2c^2 + 4c - 1)$.   53. $2\pi(R - r)$.
55. $(a - 3)(a + 3)$ 57. $(3u - 2v)(3u + 2v)$.   59. $(a - 1)(a^2 + a + 1)$.
61. $(2v + y)(4v^2 - 2vy + y^2)$.   63. $(k^2 - 2)(k^4 + 2k^2 + 4)$.
65. $(x - 2)(x - 1)$.   67. $(2k + 1)(k + 3)$.   69. $(4x + 1)^2$.
71. $(7b + 1)(b - 6)$.   73. $(4m - n)(2m + 3n)$.   75. $(6x - 1)(x + 12)$.
77. $S = 1/2(a + b + c)$.   79. $a^2 = (c + b)(c - b)$.
81. $S = 2(ab + bc + ac)$.   83. $(x - h)^2 + (y - k)^2 = R^2$.
85. $a(a - 6)(a + 5)$.   87. $2(x^2 + 1)(x^4 - x^2 + 1)$.   89. $(p^2 - 2)^2$.

EXERCISE 19

1. All integers. **3.** All even numbers. **5.** All positive real numbers. **7.** All real numbers. **9.** $\{-2\}$. **11.** $\{z \mid z < 3\}$. **13.** $\{11/9\}$. **15.** $\{w \mid w < -3/7\}$. **17.** $\{z \mid z > -1\}$. **19.** $\{x \mid x \geq 3\}$. **21.** $\{v \mid v > 1\}$.
For problems 22 to 30, let R = the set of positive rational numbers.
23. $\{v \mid v > 4 \text{ and } v \in R\}$. **25.** $\{\ \}$. **27.** $\{z \mid 0 \leq x < 2/3 \text{ and } z \in R\}$. **29.** $\{y \mid 0 \leq y < 5/6 \text{ and } y \in R\}$.
31. 1, 0, -1, -2, or equivalent answers.
33. $-1/2, -1/3, -1/4, -1/5$, etc. **35.** $\sqrt{2}, \dfrac{\sqrt{2}}{2}, \dfrac{\sqrt{2}}{3}, \dfrac{\sqrt{2}}{4}$, etc.
37. x is any positive number, and y is a negative number such that y is numerically greater than x. For example, $2 > -3, 4 < 9$.
39. All real numbers.
41. $A \cup B = \{x \mid x \text{ is any real number}\}, A \cap B = \{\ \}, \bar{A} = B, \bar{B} = A$.
43. $A \cup B = \{x \mid x \geq 1 \text{ and } x \leq 0\}, A \cap B = \{\ \}, \bar{A}\{x \mid x < 1\}, \bar{B} = \{x \mid x > 0\}$.

EXERCISE 20

1. $x < 7$. **3.** $z = 5 + n$. **5.** $b + 3$. **7.** $v = 2u - c$. **9.** $h > q$.
11. $2n$. **13.** $N - y$. **15.** $4d$. **17.** $i/12$. **19.** $1\frac{1}{2}r$. **21.** $3n/h$.
23. $25q + 10d + 5n$. **25.** 28, 31. **27.** \$23,250. **29.** 44, 46, 48.
31. $\{7/5\}$. **33.** 13. **35.** City, \$500,000; County, \$625,000; State, \$1,000,000.
37. 17 quarters, 23 dimes. **39.** 5 amperes. **41.** 400 gallons.

EXERCISE 21

1. (a) 3. (b) 6, 10, 15, 21. **3.** $m = 6, n = 15; m = 7, n = 21; m = 8, n = 28; m = 9, n = 36; m = 10, n = 45; m = 100, n = 4950$.
5. 14, 20, 27, 35.
7. From each of the m vertices, $m - 3$ diagonals can be drawn. Therefore the total number of diagonals is $m(m - 3)$. But each diagonal is drawn twice, so that the total number of distinct diagonals is $m(m - 3)/2$. **9.** 3.
11. Draw five points around a circle and connect each point with every other point.

EXERCISE 21—continued

13. The lines from the midpoints of the sides of a triangle to the opposite vertices intersect in one point. Inductive.
15. The diagonals of a rhombus bisect each other at right angles.

EXERCISE 22

1. $\overleftrightarrow{MN}$ The line has no endpoints. $\overrightarrow{MN}$ The line has one endpoint. $\overline{MN}$ The line has two endpoints.   3. Usually, the empty set.   5. Usually, the empty set.
9. True.   11. True.   13. True.   15. True.   17. No.   19. Yes.

EXERCISE 23

1. (a) Simple closed.   (b) Nonsimple closed.   (c) Simple closed.   (d) Simple.   (e) Nonsimple.   5. (a) The empty set.   (b) $\overline{BD}$.   (c) The triangle with side BD extended in both directions.   (d) The empty set.   (e) $\overrightarrow{BC}$ with B deleted.   (f) Triangle BDA with $\overline{BD}$ deleted.   (g) The region bounded by the triangle and outside the circle, excluding the boundaries.   7. The centers lie on the perpendicular bisector of the segment joining the two points.   9. Carried to 6 places, $355/113 = 3.14159$, $22/7 = 3.14286$, $\pi = 3.14159$.
11. Triangles, squares, and hexagons.

EXERCISE 24

1. Yes.   3. No.   5. No.   7. Yes.   9. Yes.   11. Yes.   13. No.
15. No.   17. Yes.   19. There are 12 triangles other than the given triangle.
21. $AFHC$ and $AGIB$.   23. None.   25. If the corresponding angles of two triangles are congruent the triangles are congruent. F. If two triangles are not congruent, their corresponding angles are not congruent. F. If the corresponding angles of two triangles are not congruent, the triangles are not congruent. T.
27. If the corresponding sides of two triangles are congruent, the triangles are congruent. T. If two triangles are not congruent, their corresponding sides are not congruent. T. If the corresponding sides of two triangles are not congruent, the triangles are not congruent. T.   29. Two to one.   31. Yes.
33. $FE = 24/7$, $DF = 48/7$.
35. If two triangles are not similar, they are not congruent. T.

EXERCISE 25

1. Two planes parallel to the given plane.   3. A circular cylindrical surface with the given line as the axis.   5. Two concentric spheres.   7. A right circular cylindrical surface.   9. A hexagonal prismatic surface.   11. Two parallel lines.   13. The interior angles of polygons of more than five sides are greater than 120°. Since at least three of the polygons must meet at every vertex, the sum of the face angles would be not less than 360°.   15. 4.

19. $V = 24, E = 36, F = 14, 24 - 36 + 14 = 2$.

21. No, since the bases may be triangles, pentagons, etc.

EXERCISE 26

1.

	a	b	c	d	e	f	g	h	i	j	k	l	m	n	o	p
Line	✓	✓		✓	✓	✓		✓	✓		✓	✓	✓		✓	✓
Point	✓		✓	✓	✓	✓	✓					✓		✓		
Rotational	✓		✓	✓	✓				✓			✓	✓	✓	✓	✓
Number of lines	1	1		2	4	2		1	3		1	6	5		3	3
Degree	180°			180°	180°	90°	180°		120°			60°	72°	180°	120°	120°

5. The student should supply the details.
Four leaved rose, point, line, and rotational symmetry.
Evolute of ellipse, point, line, and rotational symmetry.
Folium of Descartes, line symmetry.
Three leaved rose, line and rotational symmetry.
Bifolium, line symmetry.
Cardioid, line symmetry.
Elliptic paraboloid, plane symmetry.
Hyperboloid, plane symmetry, point symmetry.

EXERCISE 27

1. Is.   3. Is not.   5. Is not.   7. Is.   9. (a).   11. (b).
13. Domain and range: all real numbers.
15. Domain: all real numbers. Range: All real numbers equal to or greater than zero.
17. Domain: all real numbers equal or greater than $+2$ and equal to or less than -2. Range: All real numbers equal to or greater than zero.   19. 20.

EXERCISE 27—continued

21. 7. **23.** 69. **25.** −29. **27.** 7. **29.** −14. **31.** −17. **33.** −3.

35. $f(-x) = (-x)^3 = -x^3; f(x) = x^3$. Therefore $-f(-x) = f(x)$.

37. (a) s is the independent variable, p is the image. (b) s is the side of a square, p is the perimeter. (c) Domain and range are all positive real numbers.

39. (a) t is the independent variable, d is the image. (b) t is the time, d is the distance. (c) Domain and range are all positive real numbers.

41. (a) s is the independent variable, A is the image. (b) s is the side of a square, A is the area. (c) Domain and range are all positive real numbers and zero.

43. $f(3) = -12, f(0) = 0, f(-2) = -2$.

45. $f(3) = 36, f(0) = 0, f(-2) = -4$.

47. $f(3) = -40, f(0) = -1, f(-2) = -25$.

49. $f(3) = -6, f(0) = 3, f(-2) = -1$.

51. (a) $p = 110 + n/2$. (b) $f(26) = 123$, the blood pressure of a normal adult of age 26.

53. (a) and (b) illustrate functions, since there is but one value of y for each value of x. Figure (c) does not illustrate a function because each value of x has two values of y.

EXERCISE 28

Problem	x	−3	−2	−1	0	1	2	3	4	5
1	y	−15	−10	−5	0	5	10	15	20	25
3	y	−13	−11	−9	−7	−5	−3	−1	1	3
5	y	17	12	7	2	−3	−8	−13	−18	−23
7	y	12	7	4	3	4	7	12	19	28
9	y	−30	−16	−6	0	2	0	−6	−16	−30

11. (a)

	x	0	1	2	3	4	5
$y = 3x - 2$	y	−2	1	4	7	10	13
$y = 3x + 6$	y	6	9	12	15	18	21
$y = 3x$	y	0	3	6	9	12	15

(b) The interval is 3 in each case.

13. $v = x^3$

x	1	2	3	4	5	6	7	8	9	10
v	1	8	27	64	125	216	343	512	729	1000

15. $d = x\sqrt{2}\,d$ $\sqrt{2}$ $2\sqrt{2}$ $3\sqrt{2}$ $4\sqrt{2}$ etc.

17. $y = 3x$.   **19.** $y = -x$.   **21.** $y = 2x + 2$.   **23.** $y = 2x^2$.

EXERCISE 29

1. $(-2, -7)$, $(0, -3)$, $(3, 3)$.   3. II, III.   5. III.   7. III, IV.
9. Square.   11. Trapezoid.   13. Rectangle.
15. (b) $(-4, -2)$, $(0, 2)$, $(1, 3)$, etc. (c) The ordinate is 2 more than the abscissa. (d) $y = x + 2$. (e) No, since $-102 \neq -103 + 2$.   17. 10.   19. 5.
21. Two sides each equal $\sqrt{45}$.   23. $(2, 5)$.   25. $(1, 7)$.   27. $(-7, 1)$.
29. The abscissa of the midpoint is half the sum of the abscissas of the end points.
31. $(10, -7)$.

EXERCISE 30

19. (a) $(-5, 3)$. (b) $(5, -3)$. (c) $(-5, -3)$.
21. (a) $(-4, -7)$. (b) $(4, 7)$. (c) $(-4, 7)$.
23. (a) $(-7, 0)$. (b) $(7, 0)$. (c) $(-7, 0)$.
25. (a) $(-a, b)$. (b) $(a, -b)$. (c) $(-a, -b)$.
27. (a) (a, b). (b) $(-a, -b)$. (c) $(a, -b)$.
29. Symmetric with respect to the origin.
31. Symmetric with respect to the y axis.
33. Symmetric with respect to the y axis.   35. $(18, 0)$ $(0, -6)$.
37. $(1, 0)$ $(2, 0)$ $(-3, 0)$ $(0, 6)$.   39. None.
41. Intercept is $(0, 0)$. Symmetric with respect to the y axis. Domain: all real numbers. Range: All positive real numbers and zero.
43. Intercepts are $(1, 0)$ $(-2, 0)$ $(0, -2)$. Symmetric with respect to the line $x = -1/2$. Domain: all real numbers. Range: All real numbers equal to or greater than $-9/4$.
45. No intercepts. Symmetric with respect to the y axis. Domain: all real numbers except zero. Range: all real numbers greater than zero.

EXERCISE 31

1. Rate in miles per hour.   3. Acceleration in miles per hour per second.
5. Decibels per foot.

EXERCISE 31—continued

7. Length in centimeters per applied weight in grams.   **9.** Miles per gallon.
11. (a) $-3/2$ for each interval.   (b) y decreases as x increases.   (c) $5/2$.
13. 2.   **15.** 25.   **17.** -6.   **19.** 50.   **21.** (a) 27.   (b) 25.5.   (c) 24.3.
     (d) 24.03.   **23.** (a) $h+2$.   (b) 2.   (c) $-h+2$.   (d) 2.   **25.** -1.
27. No limit.

EXERCISE 32

1. 2, 1.   **3.** 1, 0.   **5.** $-1/4$, 9.   **13.** $y = -2x + 5$.   **15.** $y = 1/2x + 3$.
17. $y = -2$.   **19.** $y = -3/5\ x + 3$.
23. (a) $y = 3/5\ x + 3$.   (b) $y = -2/7\ x + 24/7$.   **25.** $y = -1/2\ x + 6$.

EXERCISE 33

1. (a) Rate of change is 45 miles per hour, the speed of the car.
    (b) Initial value is 360 miles, the distance from the terminal when the car started.
    (c) The zero is at $t = -8$ seconds, which is meaningless in this problem.
    (d) $f(4)$ is 540 miles, the distance from the terminal after 4 hours.
3. (a) Rate of change is -70 gallons per hour, the gasoline consumption.
    (b) Initial value is 400 gallons, the amount of gasoline at the start of the flight.
    (c) The zero is at $t = 5\frac{5}{7}$ hours, the time it takes for the tank to be empty.
    (d) $f(4) = 1200$ gallons, the amount consumed in 4 hours.
5. (a) The rate of change is -6 feet per second, the acceleration of the ball.
    (b) Initial value is 30, the velocity of the ball as it started up the plane.
    (c) The zero is at $t = 5$ seconds, the time to attain zero velocity.
    (d) $f(4) = 6$ feet per second, the velocity after 4 seconds.
7. (a) Rate of change is 5 volts per ampere.
    (b) Initial value is 0, the voltage when the amperage is 0.
    (c) The zero is at 0, the amperage for 0 voltage.
    (d) $f(4) = 20$ volts, the voltage required for 4 amperes.
9. (a) Rate of change is $9/4$ degrees Fahrenheit per degree Reaumer.
    (b) Initial value is 32, the degrees Fahrenheit corresponding to 0 degrees Reaumer.
    (c) The zero is at $-14\frac{2}{9}$ degrees Reaumer, corresponding to 0 degrees Fahrenheit.
    (d) $f(4) = 41$, the Fahrenheit degrees corresponding to 4 degrees Reaumer.

EXERCISE 33—continued

11. $-50/7$ miles per hour per second, the rate of decrease in velocity.
13. The zero is 7 seconds, the time for the car to stop.
15. $v = -50/7t + 50$.
17. Initial value is 24.00 inches, the length of the fiber under no tension.
19. $y = 0.09x + 24$.

EXERCISE 34

1. One. **3.** None. **5.** Infinitely many. **7.** None. **9.** One.
11. $(3/5, 1/5)$. **13.** $(6, 9)$. **15.** $(7, 1)$. **17.** $(4, 7)$. **19.** $(-7, 4)$.
21. No solution. **23.** $x + y - 1 + k(x - y) = 0$.
25. $x + 2y + k(y - 5) = 0$. **27.** $x - y - 9 + k(y - 8) = 0$.
29. $3y + 21x = 3$.

EXERCISE 35

1. 1250 gallons, 735 gallons. **3.** $25°, 65°$. **5.** 38.
7. \$2400 at 6%. \$1100 at 5%.
9. Rate is 9 cents per thousand cubic feet, service charge is \$1.50.
11. 13 gallons per minute and 17 gallons per minute. **13.** 7.8.
15. 25 gallons of 90% solution and 75 gallons of 70% solution.

EXERCISE 36

1. Is. **3.** Is. **5.** Is. **7.** Is. **9.** Is not. **11.** $a = 3, b = -7, c = 2$.
13. $a = m - n, b = m + n, c = 3$. **15.** $a = 3 - k, b = 7, c = k + 3$.
17. (a) $(2, -5)$. (b) Minimum. (c) $x = 2$.
19. (a) $(-3/2, -25/4)$. (b) Minimum. (c) $x = -3/2$.
21. (a) $(3/4, -7/8)$. (b) Maximum. (c) $x = 3/4$.
23. (a) $(1, 0)$. (b) $(0, 1)$. (c) $x = 1$. (d) $(2, 1)$. (e) $x = 1$.
25. (a) $(-2, -9)$. (b) $(0, -5)$. (c) $x = -2$. (d) $(-4, 5)$. (e) $x = -5, x = 1$.

Answers to Odd-Numbered Problems

EXERCISE 36—continued

27. (a) (5, 29). (b) (0, 4). (c) $x = 5$. (d) (10, 4). (e) $x = 10.4, x = -0.4$.
29. (a) (0, −9). (b) (0, −9). (c) $x = 0$. (d) Same point.
    (e) $x = 3, x = -3$.
31. (a) (3, −5/2). (b) (0, 2). (c) $x = 3$. (d) (6, 2). (e) $x = 5.2, x = 0.75$.
33. (a) (−2, 9). (b) (0, 1). (c) $x = -2$. (d) (−4, 1).
    (e) $x = 0.1, x = -4.1$.
35. (a) (3, −1). (b) (0, 8). (c) $x = 3$. (d) (6, 8). (e) $x = 2, x = 4$.
37. (a) (1/2, 7/4). (b) (0, 2). (c) $x = 1/2$. (d) (1, 2). (e) No intercepts.

EXERCISE 37

1. −3.   3. 2.   5. −3.   7. −7.   9. 0.   11. 5.   13. 8.   15. −6
17. $y = -2x + 2$.   19. $y = -4x + 3$.   21. $y = 2x - 3$.   23. (4, −4).
25. (3, 6).   27. (a) $s = -16t^2 + 160t + 40$. (b) $v = -32t + 160$.
    (c) $v_{10} = -1760$; the projectile is going down.
    (d) $v_{60} = -1920$, a meaningless answer. (e) 5 seconds. (f) 440 feet.

EXERCISE 38

1. 1, 5.   3. 6.4, 0.64.   5. 1.5, 1.   7. 3.5, −0.5.   9. No real roots.
11. 8, −8.   13. 0, 8.   15. 10, −9.   17. 5, −3/2.   19. 5/2, −1/3.
21. 3/2, 3/2.   23. 1/3, −2.   25. 1/2, 3/2.   27. 0.618, −1.618.
29. −1.577, −0.423.   31. −2.823, −0.177.   33. −1.175, 0.425.
35. $a \times b = 0 \leftrightarrow a = 0 \lor b = 0$.

EXERCISE 39

1. 3, 2.   3. 7/3, 7/3.   5. 7, −5.   7. 1/3, 0.   9. −2/3, −k/3.
11. 2/3, 1/2, −(2/3 + 1/2) = −7/6; 2/3 × 1/2 = 1/3.
13. 3.36, −0.69, −(3.36 − 0.69) = −2.67; 3.36 × −0.69 = −2.3.
15. −1.207, 0.207, −(1.207 + 0.207) = −1 − 1.207 × 0.207 = −1/4.
17. $x^2 + x - 6 = 0$.   19. $6x^2 + x - 2 = 0$.

EXERCISE 39—continued

21. $9x^2 - 3x - 2 = 0$. **23.** $x^2 - 6x + 4 = 0$. **25.** $x^2 - 10x + 23 = 0$.
27. $x^2 - 2x - 0.69 = 0$. **29.** Complex. **31.** Real and unequal.
33. Rational and equal. **35.** Complex. **37.** Complex.
39. Rational and unequal. **41.** $k = 4$ and -4. **43.** $y = 3x - 10$.
45. $y = x$. **47.** $y = -x + 11/4$. **49.** $y = -2x + 1/2$.

EXERCISE 40

1. 0.125 miles. **3.** $66\frac{1}{2}$ inches. **5.** 41.7 feet per second. **7.** 7500 items.
9. $I = E/2R$. **11.** 9/4.

EXERCISE 41

1. Sample. **3.** Sample. **5.** Census. **7.** Sample.
9. High-rise apartments are mostly occupied by small families.
11. Graduates with low incomes may not wish to reply.
13. The one machine may be the only defective machine, or it may also be the best of the ten machines.

EXERCISE 42

3. The frequencies in the intervals, from highest to lowest are: 30, 10, 20, 4, 16, 7, 4, 1, 4, 2, 2, 0. Total is 100. **5.** (a) 125. (b) 4 to 46.

EXERCISE 43

1. 41.94, 42, there is no mode, since 6 intervals have a frequency of 3.
3. 84.9, 88, 98. **5.** 85.30, 89, 98, and 99. **7.** 24.98, 24.5, 20.5.
9. $3(x_1 + x_2 + x_3 + x_4)$. **11.** $3(x_1 + x_2 + x_3 + x_4 + x_5) - 45$. **13.** 58.
15. 39.12. **17.** If there is an equal number of variates in each distribution.
19. Yes. **21.** Modal, since there must be enough cars to take care of the largest number of passengers at any time.
23. The average student, since the high or low scores have little effect on the median.

EXERCISE 44

1. 10.31, 9.19.   **3.** 13.20, 14.45.   **5.** 10.93.

EXERCISE 45

1. (a) 800. (b) 125. (c) 800. (d) 125.
3. The mean would be higher and the number of I.Q's above 115 and 130 would be considerably higher for scholarship winners.
5. (a) 25 (the mean is actually 24.98; we have used 25 to simplify the calculations). (b) 24. (c) 17 and 21. (e) 85 and 119. (f) 87 and 122. (g) The mean, median, and mode are very nearly equal, and the number of scores within one and two standard deviations from the mean is very close to 68% and 95%. (h) The distribution has two modes.

EXERCISE 47

1. $x \cdot x = x$.   **3.** $x \cdot (x + y) = x$.   **5.** $(a + b)' = a' \cdot b'$.

7.

a	b	$a \cdot b$	$(a \cdot b)'$	a'	b'	$a' + b'$
0	0	0	1	1	1	1
0	1	0	1	1	0	1
1	0	0	1	0	1	1
1	1	1	0	0	0	0

7.

a	b	a'	$a \cdot b$	$a' + a \cdot b$	$a' + b$
0	0	1	0	1	1
0	1	1	0	1	1
1	0	0	0	0	0
1	1	0	1	1	1

11. $1 + x = x + 1$, $y + x = x + y$, $0 + 1 = 1 + 0$, or equivalent answers.
13. $1 + (x \cdot y) = (1 + x) \cdot (1 + y)$ which reduces to $1 = 1$.
$0 + (1 \cdot x) = (0 + 1) \cdot (0 + x)$ which reduces to $x = x$.
$y + (x \cdot 0) = (y + x) \cdot (y + 0)$ which reduces to $y = y$, or equivalent answers.
15. $1 \cdot 0 = 0$, $x \cdot 0 = 0$, $y \cdot 0 = 0$, or equivalent answers.
17. $x + 1 = x$, $1 + 1 = 1$, $y + 1 = 1$, or equivalent answers.

EXERCISE 47—continued

19. $a' \cdot (a + b) = (a' \cdot a) + (a' \cdot b)$ by distributive axiom
$= 0 + (a' \cdot b)$ by complementation axiom
$= a' \cdot b$ by identity axiom.
21. $(x \cdot y) + z$.   **23.** $(x \cdot y) + (x \cdot z) = x \cdot (y + z)$.

EXERCISE 48

1. **3.** **5.**

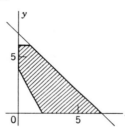
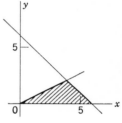

7. $(1, 0), (5, 0), (2, 6), (1, 6)$.   **9.** $(8, 3)$.   **11.** $(0, 7)$.
13. 5 units of formula 1 and 2 units of formula 2.

EXERCISE 49

1. Designate the seven members as A, B, C, D, E, F, G, respectively. Then one set of committees is $(A, B, D), (B, E, C), (C, F, A), (D, G, C), (E, A, G), (F, D, E)$, and (F, G, B). Another set is (A, B, D), (B, C, F), (A, E, C), (B, G, E), (A, G, F), (C, D, G), and (D, E, F).
5. Axiom 2.   **7.** Axiom 4.   **9.** Axiom 2.   **11.** Axiom 4.   **13.** Axiom 2.
15. Axiom 4.   **17.** Axiom 2.   **19.** Axiom 2.   **21.** Axiom 2.
23. Axiom 4.   **25.** Axiom 4.   **27.** It would be a contradiction.

EXERCISE 50

1. 21,246 years.
3. A one-to-one correspondence can be established between the sets $\{1, 2, 3, 4, 5, 6, \ldots\}$ and $\{1, 8, 27, 64, 125, 216, \ldots\}$.   **5.** $i = -50$.   **7.** $n = 36$.

EXERCISE 50—continued

9. $n = 58$.   **11.** 2/7.   **13.**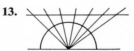

15. A one-to-one correspondence can be established between the sets $\{1, 2, 3, 4, 5\}$ and $\{1/5, 2/5, 3/5, 4/5, 5/5, \ldots\}$.

17. Each set can be placed in a one-to-one correspondence with the set of natural numbers.

Index

Abscissa, 179
Addition, of algebraic expressions, 110
    of natural numbers, 76
Aleph-null, 296
Algebra, axiomatic basis, 103
Algebraic expressions, 110
Angle, definition of, 134
    exterior points of, 136
    interior points of, 134
    measurement of, 134
    vertex of, 134
Antecedent, 28
Applications, of linear systems, 211
    of simple linear relations, 123
Arithmetic mean, characteristics of, 259
    of frequency distribution, 254
    of ungrouped data, 254
Associative property, 69, 72, 78, 107
Assumptions, basic, 17
Average rate of change, 189, 200
Axes, 179
Axioms, for algebra, 103
    for Boolean algebra, 277
    for equality, 105
    for fields, 107
    for finite geometry, 289
    of logical system, 17
    for natural numbers, 74
    Peano, 74

Bar graph, 247
Between, 133
Biconditional, 31
Binary coded systems of numeration, 64
Binary notation, 62

Binomials, 111
Boole, George, 276
Boolean algebra, 276
    application to circuits, 280
    axioms of, 277

Cantor, Georg, 294
Cardinality, 294
    of integers, 297
    of natural numbers, 296
    of rationals, 297
    of real numbers, 298
Cardinal number, 295
Cartesian coordinates, 178
Circle, 137
    chord of, 141
    circumference of, 141
    definition of, 140
    diameter of, 141
    radius of, 140
    secant of, 141
    tangent to, 141
Class, interval, 248
    limits, 248
    mark, 249
Clock numbers, system of, 67, 108
Closed curve, 138
Closure property, 67, 68, 72, 78, 107
Commutative property, 69, 72, 78, 107, 277
Complementation, 277
Compound syllogism, 10
Computers, 62
Contitional sentence, 27
    contrapositive of, 29
    converse of, 29

327

inverse of, 29
variations of, 29
Cone, 154
  lateral surface of, 154
  right circular, 154
Congruence, of angles, 145
  definition of, 146
  of geometric figures, 145
  of line segments, 145
  of triangles, 146
Conical surfaces, 154
  elements of, 154
  lateral surface of, 154
  nappes of, 154
  right circular, 154
Conic sections, 156
Consequent, 28
Constant, 119
Constraints, 284
Continuum, 299
Contrapositive, 29
Converse, 29
Convex polygon, 285
Coordinate system, 178
Curve, closed, 138
  exterior of, 138
  interior of, 138
  simple, 138
  simple closed, 138
Cylinder, 153
  bases of, 154
  right circular, 154
Cylindrical surfaces, 153

Decimal notation, 95
Decimal system of numeration, 55
Deductive system, 14
  proof, 6, 19
  reasoning, 5
    reasoning in a logical system, 14
Defined terms, 15
Definitions, special of mathematics, 17
DeMorgan's Laws, 50
Dependent variable, 169
Descartes, René, 178
Diagrams, in deductive reasoning, 8
Discriminant, 236
  applications of, 237
Distributive property, 70, 78, 107, 277
Division of algebraic expressions, 113

Domain, 169, 186, 284
Duality, 277, 291

Elements, Euclid's, 128
Equality relation, 103, 105
Equations, linear, 119
Euler's formula, 155
Exponents, 111
Exterior, of angle, 136
  of closed curve, 138

Factor, 67
  prime, 116
Factoring, 114
Field, 107
  axioms of, 107
Finite geometry, 288
  axioms of, 289
  interpretations of, 290
Finite induction, 75
Finite mathematical system, 66
Finite set, 295
Frequency, cumulative, 249
  distributions, 247
    and normal curve, 268
  polygon, 249
  table, 248
Function, 167
  definition, 168
  graph, of a function, 182
    of linear, 196
    of quadratic, 216
  initial value of, 201
  linear, 195
  quadratic, 215
  zero of, 201
Functional notation, 171

Geometric forms, in space, 153
Geometry, beginnings of, 128
  meaning of, 128
Graph, bar, 247
  of frequency distribution, 249
  of a function, 182
  of linear function, 196
  of quadratic function, 216
Great circle, 156

Hexal system, 55
  addition, 59

Index

multiplication, 59
Histogram, 249

Identity element, 67, 68, 82, 92, 107, 277
Image, 168, 171
Independent variable, 169
Induction, 75
Inductive inference, 1
Inductive reasoning, 1, 2
Inequalities, 119
Infinity, 294
Initial value, 201
Instantaneous rate of change, 191
  of quadratic function, 222
Integers, 81
  as ordered pairs, 84
  positive and negative, 84
Intercepts, 186, 197
Interior, of angle, 136
  of closed curve, 138
Inverse, additive, 84
  of conditional, 29
  elements, 107
  multiplicative, 92
  operations, 70, 77, 87
Irrational number, 94, 97
Isomorphism, 291
Iterative process, 96

Law of contradiciton, 23
Limit, 191
  of inscribed polygon, 142
Line, 133
  half, 134
  segment, 133
Linear equation, 119
  inequality, 120
  systems, 211
Linear functions, 195, 284
  graph of, 196
  initial value of, 201
  rate of change of, 200
  system of, 204
  zero of, 201
Linear programming, 283
Logically equivalent statements, 30

Mapping, 171
Mathematical system, 66
Maximum value, 284

Mean deviation, of frequency distribution, 263
  of ungrouped data, 262
Measurement, of angle, 134
  of line segment, 134
Measures, of central tendency, 252
  of dispersion, 261
Median, characteristics of, 259
  of frequency distribution, 257
  of ungrouped data, 256
Minimum value, 284
Mode, characteristics of, 259
  of frequency distribution, 259
  of ungrouped data, 258
Monomials, 111
Multiplication, of algebraic expressions, 112
  of natural numbers, 77

Nappes, 154
Natural numbers, 54, 74, 76
Necessary and sufficient conditions, 32
Necessary condition, 31
Negative integers, 84
Normal, curve, 268
  distribution, 268
  misuses of curve, 272
  properties of distribution, 269
Number, natural, 54
  system, 53, 54
Numerals, 54
Numeration, for computers, 61
  system of, 53

Octal notation, 65
Ogive, 249
One-to-one correspondence, 53, 101, 167, 295
Operations, on binary numbers, 64
  on clock numbers, 68
  on hexal numbers, 59
  on integers, 85
  on natural numbers, 76
Ordered pairs, 84, 89, 169
Ordinate, 179
Organization of data, 246
Origin, 179

Parabola, axis of symmetry of, 217
  minimum value of, 216
  vertex of, 216

Peano axioms, 74
Planes, determination of, 135
　intersection of, 130
　properties of, 135
Point, 133
Polygons, 137
　classification of, 139, 140
　convex, 140
　definition of, 139, 140
　inscribed, 141
Polyhedrons, 157
　faces of, 157
　regular, 157
Polynomials, 111
Prime factor, 116
Prisms, 158
　bases of, 158
　lateral edges of, 158
　lateral faces of, 158
Proof, deductive, 6
Proportion, 148
Pyramids, 155

Quadrants, 179
Quadratic equations, applications of, 240
　checking roots of, 235
　discriminant of, 236
　with given roots, 235
　solution, by factoring, 230
　　by formula, 231
　　by graph, 229
Quadratic formula, 231
　uses of, 233
Quadratic function, 215
　instantaneous rate of change of, 222
　maximum value of, 217, 226
　minimum value of, 217, 226
　rate of change of, 221
　sketching of, 219

Range, of data, 261
　of function, 169, 186
Rate of change, average, 189
　geometric interpretation of, 224
　instantaneous, 191
　of linear function, 200
　of quadratic function, 221
　physical interpretation of, 226
　units of, 190
Ratio, 90, 148

Rational numbers, 89
　as fractions, 90
　as ordered pairs, 89
Ray, 134
Real numbers, 94
Reciprocal, 92
Rectangular coordinates, 178
Reflexiveness, 104
Relations, 104, 167

Sampling, 244
Sections, conic, 156
　of surface, 156
Sentences, 23
　conditional, 27
Set(s), classification of, 43
　complement of, 44
　concept of, 39
　designation of, 42
　disjoint, 48
　element of, 41
　empty, 43
　equality of, 46
　finite, 41, 295
　infinite, 41, 295
　of integers, 83, 297
　intersection of, 48
　of irrational numbers, 98
　meaning of, 42
　of natural numbers, 74
　null, 43
　operations, 46
　power, 44
　of rational numbers, 92, 297
　of real numbers, 99, 297
　related, 46
　theory of, 275
　union of, 47
　unit, 43
　universal, 43
　of whole numbers, 83
Similarity, 144, 148
Similar polygons, 149
Similar triangles, 149, 150
Simple closed curve, 138
Simple curve, 138
Slope, 196
Solution set, 119, 205
Spherical surface, 155
Square root, 96

Standard deviation, of frequency distribution, 265
  of ungrouped data, 264
Statistics, use of, 243
Subsets, 43
  number of, 43
  proper, 44
  in terms of intersection, 49
  in terms of union, 49
Successors, 75
Sufficient condition, 31
Surfaces, conical, 154
  cylindrical, 153
  elements of, 154
  lateral, 154
  sections of, 156
  spherical, 155
Switching circuits, 280
Syllogism, 6
  compound, 10
  as a valid argument, 37
Symmetry, applications of, 163
  definitions of, 161, 163
  of graph, 186
  line, 160
  plane, 160
  point, 160
  as a relation, 104
  rotational, 160
System, of binary numeration, 62, 64
  of clock numbers, 67
  of decimal numeration, 55
  of geometric transformations, 73
  of hexal numeration, 55
  of integers, 81
  of linear functions, 204
  mathematical, 66
  of natural numbers, 74
  of numbers, 53
  of numeration, 53, 62
  of octal notation, 65
  of real numbers, 94

Tabular representation, 174
Tangent, to circle, 141
  to curve, 223
Tautologies, 34, 36
Transfinite numbers, 294
Transitivity, 104
Trinomials, 111
Truth, of deductive argument, 7
Truth table, for the conditional, 28
  for the contrapositive, 29
  for the converse, 29
  for the inverse, 30

Undefined terms, 15, 74, 289

Valid arguments, 34
Validity, of deductive argument, 7
Variable, 119
Vertex, of angle, 134
  of cone, 154
  of parabola, 216
  of pyramids, 155

Whole numbers, 83

Zero, 81
  of function, 201